THE GULL NEXT DOOR
A Portrait of a Misunderstood Bird

THE GULL NEXT DOOR
A Portrait of a Misunderstood Bird

Marianne Taylor

Published by Princeton University Press,
41 William Street, Princeton, New Jersey 08540
In the United Kingdom: Princeton University Press, 6 Oxford Street,
Woodstock, Oxfordshire OX20 1TR
press.princeton.edu

British Library Cataloging-in-Publication Data is available

Library of Congress Control Number 2020940432
ISBN 978-0-691-20896-1
Ebook ISBN 978-0-691-21086-5

Production and design by **WILD**NATUREPRESS Ltd., Plymouth, UK
Printed in the United States of America

10 9 8 7 6 5 4 3 2 1

Contents

Black-headed gulls

Foreword

Huge seagulls that nick your Mr Whippy are the thoughts that many members of the public have when asked to talk about this family of birds. Gulls are a bit of a dichotomy. On the one hand, they are glorious masters of the sea and air, beautiful and even occasionally colourful, with some species bearing a beautiful rosy-pink hue to their otherwise largely grey and white plumage. On the flip side, they are much-maligned members of the avian world, often vilified for assault and battery by the popular press.

Even birders have mixed feelings about them. Stop a random, binocular-wielding nature lover in the field and most will say that they give up on the gull family because the immature birds are devilishly difficult to decipher. Indeed, I know people that will actively overlook gulls for fear of getting brain-strain. Of course, at the other end of the birders' spectrum are the confirmed larophiles, or gullers, who live to count 'mirrors' on P5-P10s. You either love or hate watching them, it would seem. There appears to be no middle ground. Even the literature dedicated to them is aimed at the techie and sometimes, nerdy ornithologist.

Enter Marianne Taylor. Now, although I will have to admit that I am a massive fan of her writing, I will also have to say that at last in *The Gull Next Door,* there is now a book that bridges the gap between the nervous wannabe gull-lover and the confirmed larophile. It is an endearing, gentle and very personal account from a woman who has grown to love gulls. Within the pages of this book you will be introduced in a very relaxed and chilled way into the world of the gulls that frequent the UK. There is a tremendous spread of gull information imparted here and I guarantee that by the end you will have fallen in love.

Thank you Marianne.

David Lindo
30 April 2020

To Penny
Thank you for your words, wisdom and boundless kindness.
Squawk!

Prologue

The Old Town of Hastings sits between two steep hills. The one to the west is called the West Hill while the other, unsurprisingly, is the East Hill. The West Hill is smaller, just a big hump of mostly mown grass surrounded by streets, with a low bit of weathered sandstone cliff now well separated from the sea. This mini-cliff supports on its crown the ruins of Hastings Castle. The East Hill, its sheer cliffs slowly crumbling into the sea beyond the edge of the town, climbs steeply to its expansive, undulating summit of grassland and the gorse-cloaked 'Firehills', cut through with wooded ravines. The hilly ridge extends some 15 miles east along the coast to Fairlight, where every tree grows up leaning away from the incessant wind. Then the landscape dives down to the open floodplain of Romney Marsh, and on to Rye and the southern coast of Kent.

We lived in a very tall, thin town house on the west side of the Old Town, and from my sister's bedroom on the top floor I could see almost every house in this part of town. They were a colourful jumble of mismatched buildings – pretty Tudor cottages with doorways that even 5'3" me would have to duck to enter, blocky grey 1970s eyesores, and everything in between. They always looked to me as though they had all slid down the hillsides to land in a disorderly pile in the bottom of the valley.

I wasn't overly fascinated by this mixed-up, chaotic town, though, despite its wealth of historical interest. What caught my attention was the other town, the town-on-top-of-the-town, the town of the gulls. The Old Town had a thriving herring gull population. They built their homes on the roofs of ours, scruffy nests of dead grass stuffed in between the chimney pots, and they commuted to the beach for their daily diet of stolen fish and scavenged chips. Through the day and much of the night their voices rose above all the sounds of the street, in endless muttered and screamed conversations with partners, neighbours and rivals. By early summer there were fluffy grey chicks hatching from eggs laid in the chimney-pot nests, and soon these youngsters were out and about, pattering across the rooftops and adding their own piping voices to the general cacophony.

It only took a few weeks for that cute baby fluff to be replaced with muddy grey juvenile plumage and the sweet whistling voice to break into a sore-throated, gargling version of the adults' squawk. Each year, some of the adolescent gulls on our roof jumped before they could fly, and landed on our patio where they were at risk of fisticuffs with the family cats. As the household 'bird person', it fell to me each time to catch the errant youngster in a towel, carry it up through the house and post it out of the skylight back onto the rooftop. One day, I was engaged in this task when the young gull in my arms twisted round in its towel and lunged at my face, the blade of its bill leaving a thin but lengthy wound across my left cheek. This painful experience didn't diminish my fondness for the gulls at all but it did increase my respect for them, and thereafter I made sure the towel always covered their heads.

I would sometimes peer through the skylight at our rooftop gulls just to see what they were up to, and on one occasion I saw an adult bird that didn't look very well. Sitting on its belly with eyes half shut, it failed to respond to me tapping on the window. My gull-rescuer instincts kicked in again, and I opened the skylight and climbed out onto the narrow flat bit of the roof. The gull got to its feet with some difficulty and hobbled to the roof's edge but made no attempt to launch into flight. I caught it and brought it down into the house, where I gave it food and water. After a couple of days' TLC in the garden shed, it was walking normally, and in another day or two was beating its wings and looking fully recovered, so I let it out into the garden and it took flight. (Although I thought little of it at the time, I now find that I cannot look back on those few moments chasing down a gull on top of the roof of our six-storey house without feeling extremely sick and trembly.)

When guests came to stay, they'd stagger downstairs in the morning looking exhausted, and ask how any of us could sleep with the noise of the gulls. When I left home and went to university in Sheffield, I had the opposite problem – at first I couldn't sleep without it. I missed them and their racket terribly. I missed my daily observations of their lives, which for all the noise and bluster were more The Waltons than EastEnders – all long-lasting pair bonds, tender parenting and (largely) peaceful communal living. I missed the geography of my home town, delineated by the sea, and the seaside streets delineated by the gulls.

Leaving home brought more opportunities to birdwatch further afield, though, and so my 'gull list' grew, incorporating rarer species like the frosty white glaucous gull from the high Arctic, the incredibly dashing Mediterranean gull from southern Europe (though increasingly a fixture on British coasts), and the laughing gull (actually remarkably grumpy-looking) all the way from North America. When I got hooked on the challenge of photographing birds in flight, gulls were my favourite subjects, with their fearlessness and relaxed, elegant flight style that was easy to track with a camera. I also became aware of a sort of subculture of birdwatchers who had a special fixation on gulls – these 'gullers' or 'larophiles' thought nothing of spending whole days staking out their local landfill site, with telescope and notebook and hearts full of hope that they would find something special among the vast crowds of more common species.

I've only done a couple of hours of landfill-watching in my life, so I have to confess I'm not a true, hardcore larophile. However, my affection for gulls has only grown over the years, and my favourite gull of all is – and always shall be – the herring gull, the archetypal 'seagull', the constant companion of my childhood in a seaside town. I doubt it's a favourite for many others. In fact, this tough, glittering-eyed, loud-mouthed opportunist, with its yobbish ways and willingness to indulge in petty theft, vandalism and casual violence, may well be Britain's most hated bird. But whether hero, villain or misunderstood 'Everybird', it is the gull we all know better than any other and it is the star of this book. Here, the herring gull and its relatives around the world go head-to-head with the humans who hate and love them, in a story that spans biology and history, crime and punishment, bins and binoculars, and land, sea and sky.

CHAPTER 1

Britain's Gulls

Great black-backed gull – subadult

I'M SITTING AT my PC, and behind the open files on my screen is the image of an oil painting that I'm currently using as my desktop wallpaper. It's just something I found online and loved, and I don't know anything about it. I look it up now – it's by an Australian artist called Graham Gercken, so I guess it shows an Australian scene. Tussocky sand dunes lead us down to an idyllic-looking sunlit sea, all rendered in a loose, impressionistic style. There are no people in the image, but there are four gulls, in flight far out over the breaking waves. Except that these white shapes are not gulls really, they are just shallow, curvy V-shapes, a single brushstroke that dips in the middle. This is the universal symbol for 'gull', and it goes hand in hand with artwork depicting the sea, from children's drawings to glorious, accomplished canvasses like this one.

I don't live by the sea at the moment, but I hope to change that soon. Meanwhile, I can still go to my window and see gulls, even though I'm at least 27 kilometres from the nearest sea coast. I see them year round, but they're most numerous in winter. They'll drift over now and then in ones and twos, and later on, as dusk approaches, there will be more, in straggly little flocks. These winter gulls are making their way to some local lake where they'll roost overnight on the islands, safe from marauding foxes and grumpy fishermen. They're a mixture of species – some black-headed, some herring, some common, some lesser black-backed. That's half of all the gull species that regularly breed in Britain, ticked off in a matter of moments, though the full British gull list is much longer.

Gulls traverse the globe – there are few seasides on Earth where you won't find a gull waiting to snap up a dropped chip, Vegemite sandwich, or whatever the local delicacy might be. Although each species has its own geographic range, most of them are travellers, guided by wanderlust. Almost half of all the gull species on Earth have occurred in Britain, though most of them are rare visitors rather than regular breeding birds.

If you're a Brit, the gull you know best is probably the herring gull (*Larus argentatus*) – it's the one that steals your chips and poops on your car whenever you have a day out by the seaside. I'll be talking about it a lot more later in this book. You're also probably familiar with the black-headed gull (*Chroicocephalus ridibundus*). This small gull has a chocolate-brown (not black) hood in summer, and is a common visitor to town parks in the winter, where it entertains by deftly catching thrown chunks of bread in flight. If you're a birder, you'll probably know of quite a few other kinds of gulls, but if you're not, you probably won't – yet. So let's make a start.

Of Britain's eight regularly breeding gulls, six, including the herring gull, are very widespread, and they are all pretty easy to find and watch. The two that are not are the most recent colonists – the Mediterranean gull (*Ichthyaetus melanocephalus*), and the yellow-legged gull (*Larus michahellis*). They are both on the UK's Amber List of Birds of Conservation Concern because of their small and therefore potentially vulnerable UK populations. However, in both cases their low numbers are due to their recent arrival with us, rather than any population decline – the Mediterranean Gull is in fact increasing its population rapidly and will probably soon be removed from the list.

The Yellow-legged Gull – A Toehold in the South

The yellow-legged gull is a big, grey-winged affair. It is southern Europe's and north Africa's answer to our herring gull, and most of its breeding attempts in Britain relate to mixed pairs – a yellow-legged gull bonding with a herring gull and producing hybrid babies (or no babies). However, at least one pure pair has bred on the south coast (most often on Brownsea lagoon, Dorset) almost every year since 1995. A full-scale colonisation may happen someday, or it may not. Plenty of yellow-leggeds do come to Britain from mainland Europe after they have finished breeding, to hang out with other large gulls through the long moult season of autumn and into winter.

The yellow-legged gull was proclaimed a different species to the herring gull in 2007 by the British Ornithologists' Union, which is in charge of keeping and updating the official British List of birds recorded here and also of accepting recommended species 'splits'. Different countries have their own arbiters of these kinds of things, and decisions about what is and isn't a species are rarely universal. In the case of the yellow-legged gull, though, its separate species status was advocated by many birders long before the BOU made things official. It is consistently different in appearance and voice to the herring gull, even though those differences are small, and the two species tend not to interbreed, even where they breed together in mixed colonies in western Europe. Once the geneticists had got their hands on its DNA, they were able to confirm that, despite appearances, its closest cousin is actually the great black-backed gull, not the herring gull.

This gull can be difficult to identify, even in adult plumage. Its grey back and wings are intermediate in shade between the silvery tones of the herring gull and the charcoal-grey of the lesser black-backed, but you're more likely to muddle it up with the latter, because herring gulls have pink legs and lesser black-backs, like yellow-legged gulls, have yellow ones. It's much easier to differentiate pink from yellow than it is to assess and rank different shades of grey. There are also differences in the amount of black and white in the wingtips ... and the yellow-legged is also a distinctly bigger and stockier bird than either of the other two. This difference in size, bulk and sheer physical presence is helpful when it comes to the confusing sub-adult plumages.

The best view I ever had of a yellow-legged gull was from a stationary train on the bridge over the Thames just outside London Victoria station. I was commuting into London at the time and in winter I always made a point of checking the railings lining this bridge. Most days there would be an untidy row of black-headed gulls, and that day was no exception, but on this occasion the black-headed row had a big gap in it and in the middle of this space stood a fabulous adult yellow-legged gull, loftily eyeing its small relatives. As we sat waiting for a free platform at Victoria, I had the chance to properly scrutinise the bird at close range. A hulking male bird, it was one of the least ambiguous YLGs I've ever seen. It was robust and big-billed, with an imposing presence that recalled a great black-back, and its head was sparkly white – adult herring gulls and especially lesser black-backed gulls wear a hood of dusky streaks in winter. Its legs weren't as yellow as they would become in spring, but still plenty yellow enough for identification purposes. London is a good place to see yellow-legged gulls, but there'll be one or two in most big gull gatherings all across the south-eastern half of Great Britain, and a few more further afield – it's only really a rarity in north-west England and beyond (by which I mean further north or west than that).

A keen-eyed larophile can pick out a juvenile yellow-legged among the herring gulls and lesser black-backs at a glance by various plumage features. The breeding season is a little earlier, so the young yellow-legs look more grown-up than the rest, and begin their body moult earlier. They have whiter heads and neater black tail-bands – in fact, more contrasting plumage in general. With age, identification gets easier as the grey colour of the mantle starts to appear. If you're ever in southern France, and feeling the need to get more firmly to grips with yellow-legged gulls, take a look at any seaside spot and you should find lots of them. Do the same in Spain too, but on the southern Atlantic coast also look out for the subspecies *Larus michahellis atlantis*, a darker bird which breeds on the Azores, Madeira and the Canary Islands.

Mediterranean and black-headed gulls

The Mediterranean Gull – A Laridine Sign of the Times

The Mediterranean gull, or 'Med gull' as birders tend to call it, is a species close to my heart, because one of its British strongholds is Rye Harbour nature reserve. This south coast haven is where I cut my birding teeth as a child. It was just up the road then – today I'm living further away but I try to make the journey each spring nonetheless, just to see these beautiful gulls on their breeding grounds.

This part of the south coast, straddling the Kent/East Sussex border, is all about shingly beaches. Dungeness is the largest shingle peninsula in Britain, and Rye Harbour is another shingly wilderness just a few miles west. They give you an idea of what a shingle beach can be like if it's left to nature rather than handed over to human holidaymakers – the further you go from the sea, the more lush and complex the shingle flora becomes. Lovely flowering plants like the intensely blue-violet viper's-bugloss and the showy yellow horned-poppy grow here, and birds like ringed plovers hide their eggs among the pebbles.

Rye Harbour nature reserve has several lagoons dug into the coastal shingle, with islands for birds to nest on, and hides for birders to watch the birds. In spring, the islands fill up with nesting black-headed gulls.

You can hear them from a good way off as you walk through this bleak, windy landscape. They're raising such a racket that you wonder whether your ears will cope when you are close enough to actually see them. The yelps of herring gulls sound positively musical compared to the shrill, nerve-grating screeches of all those black-headed gulls. They need to put themselves forward in this way, though – a shy and unassuming black-headed gull won't do too well in life. They are competing with each other for space on the islands – each pair must commandeer enough room for a simple pile-of-straw nest. And they're also working together to protect the whole colony. They're not the biggest of gulls, so need each other's support to make a real show of strength.

When I go to see the Med gulls that nest here alongside the black-headeds, I have to listen for a different voice. The beloved writer Richard Adams introduced a black-headed gull character who befriended the rabbits in *Watership Down*, and called him 'Kehaar' – a fair approximation of the black-headed gull's nerve-shredding scream. (I talk about Kehaar some more in Chapter 5.) If Richard Adams had gone with a Med gull instead, the name would have been something like 'Yeowk', but no one word can convey the fruity, interrogative pitch to the Med gull's distinctive call. There's a distinct, fruity Kenneth Williams timbre, which cuts across the black-headed gulls' wall of screaming sound. When I search for the bird that's saying 'yeowk' among that vast confusion of shrieking, wheeling shapes, I'm looking for a pair of white wings, without the black trailing edge of the black-headeds. And there it is – a shade bigger and a whole lot burlier, with a scarlet bill, big white eyelids and a jet-black hood that's pulled well down its neck instead of riding up on the nape like the black-headed's (whose hood is also actually brown). And I'm especially enchanted by those beautiful snowy wings, the slim bones of flight showing through clearly when the sunlight is behind them. The Med gull lands dead centre on an island, and its mate greets it with an upraised head. These two have occupied the best spot to nest, pushing their smaller cousins out of the way, but they will join in the team effort when danger threatens. After watching the colony for a while, I head off down a gravelly path on the way to another shingle-edged lake where the gulls go to bathe, and watch the Med gulls sailing overhead on their way back from a bath, looking pristine and uttering the occasional saucy 'yeowk' as they go.

You might hear a 'yeowk' in quite a few other places too these days, some of them surprising. I heard it while out on the North Downs the summer before last, looking for Adonis blue butterflies on a fragrant flowery hillside somewhere not far from Ashford but a good long way from the coast. I looked up to see several angelically white-winged Med gulls floating over at some considerable height. They were incongruous in this landscape of rolling green, but apparently this inland passage of Mediterranean gulls is a regular thing. I've also heard a surprise 'yeowk' while exploring the Wildfowl and Wetlands Trust's reserve at Barnes in west London, and up north(ish) in Cheshire at the delightful Woolston Eyes nature reserve, near Warrington. Here, black-headed gulls nest on islands in shallow marshy waters, and from a high tower hide I saw a pair of Med gulls among them.

This is how Med gulls first colonised southern England — a pair or two breeding within a black-headed gull colony, and numbers gradually growing and sending out pioneers to find new colonies to join. The first recorded breeding took place in 1968 in Hampshire. In the early days of the Med colonisation there were, as with the yellow-legged gull, a number of mixed pairs as the few Med gulls around failed to find each other, and wooed a black-headed gull mate instead. Now in 2017 there are about 800 pairs in Britain, mainly in the south and south-east, but spreading north. The first Yorkshire breeding record happened in 2010 at RSPB Old Moor — a site that is not only decidedly northerly but well inland. A few Med gulls have also oversummered at sites in Scotland, batting their impressive eyelids at black-headed gulls and, in some cases, forming mixed pairs.

In winter, Mediterranean gulls leave their breeding colonies and wander about. Like black-headed gulls, they lose most of their winter hoods after breeding, but the winter adults at least are still easy to identify with their pure white wings. The winter head pattern of both adults and youngsters differs from that of the black-headed gull too.

Where the black-headed has a neat dark spot behind the eye, the Med has a blurred black bruise-like smudge that extends just in front of the eye as well as behind, making it look a lot grumpier and more pugnacious than its smaller cousin.

These days, Med gulls are really quite widespread in winter, especially along the south coast, as climate change makes our country ever more hospitable to them. You'll see them at many seasides in Kent and Dorset in particular. I spent a very happy afternoon in March 2017 photographing several of them at Herne Bay on the north Kent coast. The following January I met a few at Weymouth, in Dorset, and not long after that I had a coffee at the end of the harbour arm in Folkstone while watching several lovely Meds wheeling over the calm harbour waters. They are also frequent at some places on the coast in East Anglia, east Yorkshire, on the Wirral coast, and even in eastern Ireland. It's not just here that this gull is thriving, either – it is also spreading along the north coast of Europe and now even breeds in Sweden. A lot of ringed Med gulls wintering in the UK are breeding birds from northern and eastern Europe, for example the Netherlands, Germany, Hungary, Poland. With each year that passes, the Mediterranean gull's name becomes more inappropriate.

The Black-headed Gull – Is it having a Laugh?

That brings us to our six more established and more widespread breeding gull species, starting with the one that helped the Mediterranean gull to establish itself here. The black-headed gull (*Chroicocephalus ridibundus*) is a very familiar bird to most of us, in both town and country. We have about 140,000 breeding pairs in the UK, and in winter numbers swell to 2.2 million individuals, as breeding birds from Europe and beyond head over to enjoy our slightly milder winter climate. So, even in years when our breeding pairs produce lots of young, the great majority of black-headed gulls here in winter are from overseas, with those from the nearer countries arriving in autumn, and those from further east, as far as Russia, not turning up until midwinter.

It's December now as I write, and I know that if I head down to the local park I'll see lots of them around the lakes, rubbing shoulders with the assorted formerly domestic mallards and muscovy ducks that have been dumped there by disenchanted backyard poultry farmers. In fact, I don't even need to go out. Right now, just a quick look out of the back window across the northern edge of town will reveal a few flying past, lit from below by the gathering snow on the rooftops. They are graceful, highly agile in flight, and their slim, pointed wings have a broad, white leading edge and a narrow, black trailing edge, with light grey in between. No other resident British gull shows this wing pattern, so it's always easy to identify a flying black-headed gull when you get a good look at its opened wing – the rule works regardless of age and it applies at any time of year.

Black-headed gull

When perched, the black-headed gull might be very slightly trickier to identify. It is smaller than all our other resident gulls – it's just a shade bigger than a pigeon, and has a skinny, gangly look. Its long wings show black at the tip, with tiny white spots at the very tips of the feathers when it has fresh plumage. The grey of its back and the inner parts of its wings is paler than that of a herring gull, its eyes are mild and dark, and its bill and legs are quite skinny. It has, as

I said before, a shrill and grating screech of a voice, though it's not at all unpleasant to hear when it's just a bird or two calling, rather than a whole breeding colony. Nevertheless, whoever decided that *ridibundus* (laughing) was a good species name for this bird must have been more of a guffawer than a chuckler.

Down at the park, the black-headed gulls wheel about over the water, and sit on top of the short posts that border the lake. Often, one will fly in and, rather than settling on a vacant post, will head for an occupied one and kick or shoulder-barge the gull already there, forcing it off, with much shrieking from both parties. There are adults and first-winter birds here, the latter distinguishable by the brown speckling on their wing-covert feathers, their dark tail-band, and their orange rather than bright red legs and bills. The adults' dark hoods, a marker of breeding condition, are replaced with white for the winter, with just a round dark smudge behind the eye and another, hazier smudge that goes up from the eye and over the top of the crown. The white eye-ring that's prominent in the dark hood is not prominent at all in winter but with a close look you can see it, contrasting with the dusky head-smudge. A few particularly hormonally progressive individuals are already sprouting quite a few extra dark head feathers too – the beginnings of next year's hoods, even though it's still only December. This is why it's a bit inaccurate to refer to 'summer' and 'winter' plumage when talking about adult gulls – 'breeding' and 'non-breeding' is better. (In America, they go for 'basic' (for non-breeding) and 'alternate' (for breeding) – but I plan to stick with breeding and non-breeding.)

Black-headed gulls, like most small species, are 'two-year' gulls. This means that they have fully adult plumage by the time they are two years old. This makes life easy for the gull watcher. All the birds I can see are either in their first winter, or they're adults. The adults could be anything up to 30 years old (the European age record for a ringed black-headed gull is 30 years and 7 months). If I could catch them and minutely examine them I might be able to pick out the ones that were in their second winter – maybe the odd retained juvenile feather. But otherwise there's no way to tell how old they are, unless they happen to be ringed. The same goes for sex – males and females are almost identical. As with other gulls, the males average a shade larger and heavier, with a bigger bill and more angular-looking head. If you see a pair together you can often make a fair guess at which is which. But in a big jumble of birds, picking the boys from the girls is a lot more difficult.

The black-headed gulls in the park in winter don't breed here, or anywhere near here. Breeding is not on their minds – eating is. When a human family arrives, stops at the lake shore and someone pulls out a bag of bread crusts, the gulls go berserk. The humans are more interested in feeding the cute ducks than the noisy gulls – at first. But then one of them throws a bit of bread into the air, into the circling gulls, just to see what will happen. One deftly catches it, and gulps it down even as all the others nearby launch themselves, screaming, towards the prize. It's fun to play catch with the gulls. Bread is, as we are often told these days, not the best food for ducks and not for gulls either, but scroungers can't be choosers – free and easy food like this is one of the reasons why so many black-headed gulls become urbanites for the winter.

In spring and summer, the adult black-headed gulls will leave places like my local park, and assemble at the lakes and marshes and shingle shores, where they breed in large, noisy colonies. The youngsters often hang on at the parks well into spring, but sooner or later they usually head for the breeding colonies as well. They probably won't nest, but they might experiment with courtship behaviour, learn the lie of the land for next season, and generally get in the way of the adult breeding birds and make a nuisance of themselves. I once watched a pair of first-summer black-headed gulls within a small breeding colony, which had clearly fallen for each other and were defending a little territory on the island, although they had not built a nest. It was quite early in the season and the adult pairs around them were mating frequently, but when these two tried, they couldn't seem to get it right. The male jumped on board the wrong way round, then stepped on the female's head and made her tip over sideways. They gave up and stood side by side, looking dejected and – to my imagination, discreetly watching their more experienced neighbours for tips and tricks. Apparently, even this most primal of acts can't be guided entirely by instinct.

There's great variation in how quickly black-headed gulls become black-headed, or brown-hooded if we're going to be really accurate. While some have gained the best part of their hood by mid-January, most take a bit longer and are not fully hooded until March or April. Black-headed gulls in their first summer do develop a full hood, though later than the adult birds do, and look strikingly variegated with their retained dark wing and tail markings. They replace these juvenile feathers in their first full moult, at a little over a year old. Fledgling juvenile black-headed gulls are quite dramatically different from adults – their first plumage has lots of gingery-brown in it and confuses many a beginner birder. However, they can usually be recognised by the fact that they are following adult black-headed gulls around and screaming at them. They have usually moulted most of their juvenile body feathers by the time they move away from the breeding colony, and are in their much whiter, more adult-like first-winter plumage by the time they are fending for themselves and showing up around the park boating lake and duck pond.

The black-headed gull is on the Amber List of Birds of Conservation Concern in the UK. This designation is primarily because of recent declines in its wintering population here – it is declining moderately as a breeding bird in several northern European countries. Its UK breeding population has remained broadly stable since the 1960s and 1970s. However, a couple of hundred years ago this species was a *bona fide* rarity in Britain – it only really became numerous and started to establish very large colonies in the 20th century. This was quite a boon for the people of Britain. In our hunter-gatherer days we humans regularly raided birds' nests for their eggs, a useful source of protein – but these huge and accessible black-headed gull colonies were exploited by people on a near-industrial scale through the 1930s and 1940s. Some 300,000 black-headed gull eggs were sold at London's Leadenhall Market each year through the 1930s.

All birds' nests in the UK today are protected by law, with just a few exceptions. The trade in black-headed gull eggs is one such exception – it remains legal today but is strictly regulated. There are only about 25 licensed collectors and they can each only collect a limited number of eggs, with everything meticulously documented. In 2015, 27,842 black-headed gull eggs were sold via fancy food outlets and restaurants, for up to £10 apiece.

While I believe in doing your research, I draw the line at eating a black-headed gull egg to tell you what they're like. I don't even much like hen's eggs, after all. However, I did read up on what other people thought of them. Apparently, they are particularly rich and creamy with an intensely red-gold yolk, and are best served soft-boiled. This serving method also means that you can leave a bit of the shell on for dramatic effect and make it clear to your guests that this is no ordinary egg. Like all gulls' eggs, they are quite beautiful, with blue-green shells marked liberally with spots and splotches.

The Not-so-common Common Gull

When you say 'common gull' to a non-birder, you may want to clarify by explaining that you mean the species called 'common gull', *Larus canus*, and not just any old gull that's common as muck (which in most people's minds would be the herring gull, *Larus argentatus*). The common gull is, as nearly every British birder will eagerly tell you, not actually as common as you might think. The UK breeding population is 49,000 pairs, which sounds like loads but is only a third as many as we have herring gulls and black-headed gulls. In winter, as with black-headed gulls, more arrive from the continent and we then have a total of some 710,000 individuals in the UK.

This is a middle-sized gull species, in between black-headed and herring gull. It's rather elegant and pretty, with long, darkish-grey wings, tipped black with white spots, and a cute, round head with non-threatening, big, dark eyes. It has greeny-yellow legs, and a relatively dainty yellow bill which lacks the red spot near the tip that you'll see on the large gull species, but does develop a smudgy-looking dark ring in winter. Its voice doesn't match its quite delicate appearance. In America, they call this bird the 'mew gull', for its grating, drawn-out call, which recalls the cry of a disgruntled and sore-throated feline. Before I heard one calling, I used to think that its species name, *canus*, might have something to do with *canis*, Latin for dog, but then found out that *canus* is just Latin for 'grey' – these gulls caterwaul, they don't yelp.

When I think of common gulls, the first image that comes to mind is the view I have enjoyed from the overnight 13-hour London–Inverness coach, heading up the A9 towards Pitlochry and Aviemore on a morning in early spring. It's a killer of a journey, but a keenly priced killer – ideal for the impoverished writer who lives in Kent but needs to be in north Scotland. By the time you reach this point you've got through a long night trundling up the M6 and you're feeling quietly self-congratulatory at having survived those first nine hours. You're still cramped and exhausted but the sun has risen and you are being rewarded with some really stunning views across the Tay and Tummel, and the hills beyond. Down along the river fly white birds, too distant to make out any detail but there is something very floaty and angelic about their leisurely progress. They are common gulls, prospecting for nest sites on the shingly shoals and islands. Scottish rivers and lochs are important breeding sites for this species – in my neck of the woods very few breed at all and for us, they are winter birds.

If your local park attracts lots of black-headed gulls in winter, there's a fair chance there will be some common gulls among them too. The adults are less pretty in winter plumage – like many other gulls they develop a hood of grey streaks, and their bills and legs become drab, the latter marked with a blurry dark ring. There may well be youngsters present too.

The common gull is a 'three-year gull', taking one year longer to reach adult plumage than the black-headed, so there are two different categories of young 'uns to look for. There are the first-winters with their mostly mottled grey-brown plumage, relieved only by the solid grey back (or 'mantle', or 'saddle'). They have broad, dark wingtips and a dark band at the end of the tail. Then there are the much more adult-like second-winter birds. These still show a dark tail-band, a bit of brown in their wings, and a few other tell-tale signs of their youth.

Those in their third winter look just like those in their fourth, fifth or thirtieth year. There is in fact a Danish record of a common gull observed alive and well thirty-three years and eight months after it was ringed. Among European gulls, this record has been beaten only by herring and lesser black-backed gulls, both of which are considerably bigger and tougher birds.

I like to visit the central London parks on sunny winter days to take bird photos, and one of my top targets is the common gull. I think of it as a little bit more exotic than the herring and black-headed gulls that are there too – probably because the common gull wasn't a bird I saw very often through my south-coast upbringing. But London delivers these lovely gulls in spades. The Round Pond in Kensington Gardens looks about as unpromising as any body of water could from a wildlife point of view. It's a biggish, round lake that's encircled by pavement – no emergent or marginal vegetation. Yet there are always lots of birds on it in winter – mute swans, greylag geese, a grebe or two, tufted ducks, shovelers. One time, there was a drake scaup swimming in small circles in the centre of the pond, looking bewildered – and no wonder, as the scaup is a seafaring duck, an uncommon visitor anywhere inland, let alone the heart of London. There are also gulls aplenty. I always look out for the common gulls first – the adults dapper despite their speckly winter hoods, the young ones pleasingly mottled but with neat, grey saddles. They have a rangy, sleek look and a fearless manner. When one grabs a thrown ham sandwich and takes to the air, a couple of herring gulls pursue it but they can't outmanoeuvre the common gull. It twists and turns, dives like a falcon and scythes across the water, leaving its pursuers flailing, all the while gulping down its prize with indecorous relish.

When I've visited north Scotland in summer, I have spent some time with common gulls in breeding mode. At Loch Maree near the west coast, they nest in small groups on the islands, alongside that region's glamorous and rare speciality, the black-throated diver. On Mull I ate my sandwiches beside a beautiful sea loch with seaweedy islands, on which immaculate summer-plumaged common gulls pottered and bickered. And at Dungeness I've seen them snoozing in pairs on the visitor centre roof, but I've never been in the right place at the right time to get a proper first-hand insight into their family life.

The common gull, like the black-headed gull, is an Amber-listed species on the UK list of Birds of Conservation Concern. This is because of declines in the breeding and wintering population. Some Scottish regions in particular have shown major declines in breeding numbers, and breeding populations are declining in north-west European countries too, but this gull is not particularly well served in the UK or abroad by surveys at present.

The Kittiwake – Sentinel of the Seabird City

The other 'middle-sized' gull to breed in Britain is the kittiwake (*Rissa tridactyla*). 'That's a funny name,' you might think. 'Why's it not called the "something gull" like all the others?' Indeed, the kittiwake and its Pacific cousin the red-legged kittiwake, the two members of the genus *Rissa*, are the only gulls in the world that aren't called 'something gulls' in English. This leads to careless newspaper journalists saying things like 'kittiwakes should not be confused with seagulls', but they are 100% gull. Kittiwake is onomatopoeic – it describes the harsh, rather donkey-like creaky and screechy call these birds give on their breeding grounds. Approach a busy seabird cliff in summer and you'll quickly realise that the wall of noise that emanates from it is mostly from kittiwakes, ceaselessly yelling their own name above the rumbling waves.

Kittiwakes' close connection to coasts with steep cliffs means they're probably the least familiar of our breeding gull species, but they're also the most numerous, with 380,000 pairs in the UK (more than double our breeding population of herring gulls). Most of these breeding birds, though, are living on remote Scottish headlands and rugged islands. Westray island in Orkney has nearly 40,000 pairs, for example. The further south you go, the scarcer they become.

An adult kittiwake in breeding plumage looks a lot like a common gull in breeding plumage. If you look at their faces only, they are close to identical. I once made a composite photo of the faces of a kittiwake and a common gull, and asked a few birding friends to tell me which was which – they found it surprisingly tricky. Even if you can see the whole bird, the two species are still pretty similar. They're about the same size, they are both smart-looking grey-winged gulls with big, dark eyes and pure yellow bills. The most obvious differences, at first glance, are that the kittiwake has solid black wingtips, while the common gull's wingtips are black with white spots, and also the kittiwake has noticeably short black or occasionally dark-reddish legs, while the common gull's are longer and greeny-yellow.

The Atlantic subspecies of kittiwake has unusual foot anatomy to go with its stumpy legs in that it usually has no hind toe (hence *tridactyla* – 'three-toed'), while common gulls and other gulls do have a small but fully formed hind toe at the back of their foot, as well as the three full-length and webbed ones at the front. (The Pacific subspecies of kittiwake, *Rissa tridactyla pollicaris*, has a hind toe too – *pollicaris* comes from the Latin word *pollex*, meaning 'thumb'.) Kittiwakes also look a little stockier and more robust in the body. The other differences, though, are to do with the way they move and behave, and these become more and more obvious the more time you spend watching kittiwakes and common gulls in action. Eventually you'll wonder why you ever thought they were similar at all.

Kittiwakes don't really 'go' with other gulls. For wildlife watchers, they belong with different birds – the auks, gannets and shags that share their vast, deafening, rowdy cliff-face colonies and open-sea foraging grounds. However, there are some places that hold colonies of just kittiwakes, one of which is not far from where I live. Seaford is a pleasant small town on the south coast, not far from Eastbourne, and within walking distance of some truly stunning coastal scenery. Head east from the town seafront and you'll soon be climbing up the very steep grassy bank of Seaford Head, with white cliffs dropping away to your right. From the high tops of the head you'll see down into the exquisite Cuckmere Valley with its meandering oxbow lakes – this is the only significant river valley on England's south coast that is still free of urban development. However, you don't have to climb that high before you see a huge lump of chalk, slightly separated from the main cliff face and lacking a grassy crown. Here (in summer) you'll find Britain's southernmost breeding colony of kittiwakes, and there are a few spots where you can watch them without going too near the cliff edge (but *please* still be careful!).

There will be herring gulls sailing past, and quite possibly fulmars too, the latter resembling mini-albatrosses with their scowls and complicated bill structure, but with deceptively gull-like coloration. The kittiwakes are easy to pick out from these imposters – their beauty and their light elegance on those long, slim wings is several orders of magnitude above the others. Their white bits seem whiter, their black bits blacker, their grey bits not so much grey as silver-blue, with a striking and lovely translucence to the flight feathers when the light hits them just right. If you visit in summer you'll see chicks in the nest, and by late summer you might also see these young birds of the year flying about, getting used to their new wings. Here is where kittiwakes really differ from our other gulls. In juvenile plumage they are not a speckled muddy grey-brown, but incredibly dapper in black, white and grey. You'll notice a black neck-boa, cheek-spot, zigzag wing-stripe and tail-band. The wings are grey where they are not black, and the rest of the bird is pearly white. In winter, adult kittiwakes look a bit like their younger-selves, donning a neat, soft grey neck-boa and a dusky cheek-spot, rather like that of a winter-plumaged black-headed gull. They also have a tiny touch of grey around the eye, lending their expression a slight frown of dissatisfaction. At this time of year, you can tell them from common gulls (with their streaky grey winter hoods) at a glance.

Even newly hatched kittiwakes look different to very young gulls of other species – their fluff is whitish rather than drab mottled brown. The bold beauty of the nesting and adolescent kittiwake reflects its upbringing, on a tiny ledge on a sheer cliff face where camouflage is really not needed. It also reflects how the young bird will be living once it leaves the nest. Not for it a life of loitering on fishing beaches or city parks – it will be roaming the high seas, and its zig-zagging upperside pattern does in fact give it camouflage when seen from above, flying over a rippling sea. A similar pattern is present in a range of other ocean wanderers, such as the prions, the soft-plumaged petrels, and some of the other more seafaring gulls.

I've seen kittiwakes at their nests many a time – most memorably on the Farne Islands in Northumberland, where you can watch at point-blank range as they go about their business. They are charming to watch in this way – tender towards their mates and chicks, and endearingly ungainly on their short little legs. But my favourite encounters have been at sea.

Watching birds from boats is one of my favourite ways to spend time, but the kittiwake is the only gull that really features on a typical 'pelagic' trip around the British Isles once you're out of the harbour and heading out to sea. As the land recedes, you leave behind the gatherings of herring and great black-backed gulls, and you start to see gannets, huge, graceful and brilliant white against the moody, surging sea. You also see auks bobbing on the waves and clattering away in a frantic panic if your path takes you too close – they look like skimmed stones on wings, barely clear of the crests. Fulmars are not afraid at all and cruise on flat wings alongside the deck almost at touching distance, keeping pace with the boat. At this range, their intently frowning faces and weird, complicated bills clearly reveal their allegiance to albatrosses, despite their superficial likeness to gulls. Further out, their relatives, the Manx shearwaters, might flash alongside for a little while, flying fast and confidently over and in between the shapes of the shifting swell. The kittiwakes fly closer and higher than this, the light catching the translucent silver of their flight feathers. They have a stylish beauty and grace that sets them apart from other gulls – the shapes they throw on the wing, the crisp clarity of their markings, and their aura of ease over even the heaviest of seas – these things combine to create an impression of a true, consummate seafaring bird.

Kittiwakes are on the Red List of Birds of Conservation Concern in the UK. Their numbers in Britain have fallen alarmingly, and the same goes for them in other parts of the world. This is the only one of our breeding gulls to have a global status of Vulnerable (as assigned by the International Union for the Conservation of Nature, aka IUCN). All of our other gulls are categorised as Least Concern, meaning that they're not at risk of global extinction any time soon. Species classed as Vulnerable, though, are at real risk, and are likely to be reclassified as Endangered if their circumstances don't improve quickly and significantly. Endangered and Critically Endangered species are those that are considered to be at serious risk of extinction in the near future.

The kittiwake is our most pelagic gull and, perhaps, the one least likely to be able to adapt to changing conditions. The circumstance that's particularly driving their decline is climate change – sea temperatures are rising and that's causing plankton populations to crash. Small fish and other marine animals feed on plankton, so their numbers are falling too, and the disaster spreads along the food chain. The problem is apparently worst in winter.

That leaves three big gull species that are widespread breeding birds in Britain – the herring gull, the lesser black-backed gull and the great black-backed gull. The first of these is getting a whole chapter to itself, so let's look at the two 'black-backs'. First of all, which one's which? The clue is in the name; one's lesser and the other is great. Really great. The greatest of all the gulls, no less.

The Great Black-backed Gull – Majesty and Might

When you see a great black-back close up you are left in no doubt that this is one formidable feathery beast. It is huge, and even given its general hugeness, the particular hugeness of its bill is an arresting detail. A bill the size of a family hatchback, to use the memorable description offered by a birding friend of mine. The smallest female great black-back might not stand out *that* much alongside other gulls, but she'll probably still catch your eye, and her towering hulk of a boyfriend certainly will.

Great black-backs, like other large *Larus* gulls, take four years to reach adult plumage. Those in their first winter are the familiar dappled grey-brown of most baby gulls, but they have rather white heads and a generally fairly contrasting look compared to other species. Plus, of course, they are huge. The same goes for second-winters and third-winters as their black upperside plumage begins to appear, at first in the middle of their backs on the mantle, and then gradually over the inner parts of the wing. By the fourth winter, most look like adults, with just the odd brown-tipped tail feather to give them away.

So we come to the adult great black-back. It is a vast, barrel-shaped monster, with proportionately shortish wings. Its head, body and tail are snowy white. Its back and wings are coal-black. The wingtips are perhaps a shade blacker than the other bits but there is no obvious contrast. The outermost flight feathers have a bit of white in them near the tips. This is the norm for adults of the big gulls, and the amount and arrangement of white bits is key for identification in some cases. Start looking into this, and you'll quickly find yourself lost in a tumult of jargon which I'm going to have to explain sooner or later, but for now let's keep things simple. You can tell the great black-backed from the lesser by the fact that the greater's outermost flight feather has a big, solid white tip. And also by the blackness (rather than dark greyness) of its back, the fact that it has pink rather than yellowy legs, and, most importantly, by the fact that it's huge.

Great black-backs are proper 'seagulls' – it even says so in their name, *Larus marinus*. Gull of the sea. They are the rarest of all our resident gulls, with a breeding population of just 17,000 pairs. In winter there are more of them here – some 76,000 individuals which mainly come our way from northern and north-eastern European coasts. However, even in winter they don't turn up at inland roosting spots nearly as much as other gulls do – they stick to the coasts. You'll find lots of them around fishing harbours. The fishermens' beach at Hastings, where the off-duty boats sleep amid the net piles and strandline debris, is a top hangout for them. They're outnumbered by the herring gulls, but outclass them by a mile. When I'm back in my home town, I go to visit them and watch from a respectful distance as they loaf on the shingle or stride about looking for food, effortlessly barging the herring gulls aside to secure the best pickings.

Another good place to commune with the king of gulls is down on the western tip of Cornwall. On the harbour arm in Mousehole village, I came face to face with a big male great black-back. I was on the footpath below, he was up on the harbour wall, so I could look right into his little grey-green, seawater-tinted eye with its scarlet ring of skin. When I raised my camera, his head was frame-filling. Later, I took a walk up to the bird hospital on the hill out of the village. Here, locals can bring any injured wild birds they find – the hospital patients are cared for until they can be released, and those that are permanently disabled to the point where release is impossible are cared for indefinitely. In a spacious aviary of non-flying gulls, several herring gulls were standing around (or 'loafing' as some of the field guides put it) by their pool, much as they would do in the wild, and with them was a single adult great black-back. Like the free-living one I'd been admiring at the harbour, this one was impressively big, and quiet and implacable in its manner. It met my gaze with a steady stare of its own. Its bill was a truly fearsome thing to behold, and I had no doubt that, come feeding time, it would have no difficulty intimidating its way to the choicest portions. But right now it wielded its weapon with noble calm.

Less noble were the antics of the great black-backs in Newlyn Harbour, a mile or two down the road. Among the colourful boats floated many gulls in search of scraps, and the mood among them was cantankerous. I photographed several disputes, including a fierce confrontation between an adult and a first-winter great black-back. The adult was vigorously disciplining what may or may not have been its own offspring with a relentless volley of pecks and feather-yanks, while the distressed youngster yelled and paddled in circles, unable to gain the space it needed to take off or fight back.

I also have fond memories of watching great black-backs from the towering clifftops of Troup Head, in Aberdeenshire, on an intensely sunny June afternoon. These steep rocky faces are rammed solid with seabirds through summer. The place is full-on sensory overload with sea smells and guano smells, plus the din of bawling kittiwakes, growling guillemots, clattering gannets, and crashing waves down below. All these birds were sailing or frantically flapping to and fro non-stop just under eye level. Every now and then a lone great black-back would go by, on the hunt for a nest to plunder, and maybe I imagined it but it seemed the noise quietened a little at its approach, the other birds discomfited by its presence. Its black-and-whiteness against the deep indigo swirl below was glorious, but the sweep of its gaze across the nesting birds, each of them vulnerable to its predatory whims, was a little chilling. A great black-back can, if it wishes, gulp down a small seabird like a petrel whole, and can hatchet up larger items with sheer blunt trauma applied through that tremendous bill.

Lesser black-backed gull

The Lesser Black-backed Gull – Sleek and Stylish

The lesser black-backed gull really doesn't have much in common with the great, beyond sharing a genus and a general colour scheme. This is an elegant bird, slimmer-bodied, longer-winged and altogether more gracile even than a herring gull, let alone a great black-back. In the UK we have about 110,000 breeding pairs, and fewer in winter (130,000 individuals) because although some come here from Europe, many of our breeding birds migrate south, most of them passing the colder months along the Atlantic coast of Spain and north Africa. However, there's a change afoot – more and more lesser black-backs now opt to stay put, and they are a key component of the large flocks that visit reservoirs and rubbish tips all through autumn and winter.

They're also turning up more and more in urban parkland, where they make a living mainly from human leftovers but supplemented with the odd surprise. I know I've just said how terrifyingly predatory the great black-back can be, but it was a lesser black-back that was caught on camera in 2014, drowning and eating a pigeon in Hyde Park, London. The press jumped on this story with predictable glee, and found that the attack wasn't a one-off – birders who visited the park had seen the same gull doing the same thing regularly for at least five years.

I found a video of this menacing bird in action. Its attempts to creep up on a pigeon looked laughably bad at first, because it was doing so on a large open pathway, and despite it adopting a low, stealthy slink, its approach was very obvious. However, pigeons preoccupied with breadcrumbs are not that attentive and after a few failed pounces the gull did manage to seize a pigeon in its bill, and immediately took it out to the lake to try to drown it. The pigeon struggled free but not before the gull had divested it of some of its plumage. A second victim didn't manage to get away. After drowning it in the lake, the gull carried the dead pigeon back to dry land and laboriously shook and tore its body apart, while other lesser black-backs gathered around to try to grab a share.

Lesser black-backs may not be the most ferocious of the large larids but they are just as enterprising and, when necessary, predatory as any. In their social habits they're quite similar to herring gulls, though more likely to nest on flat ground, such as grassy islands in rivers and lakes, than on cliff ledges, more likely to breed inland, and more colonial too. They will nest on town rooftops, sometimes in large numbers if they find a big, flat factory roof or similar that they can pretend is an island in the urban jungle. However, they are not anything like as much a feature of the seaside town scene as the herring gull, at least not in the seaside towns that I know well. Hastings seems to have only a handful of pairs. I'm not sure whether herring gulls would outcompete them for nest sites (they are a bit smaller and lighter-weight than herring gulls after all), or whether the more restricted nest spaces available on small house roofs just don't suit them.

Little gull

Gulls on Tour – Visitors to Britain

Besides our breeding gulls, we have several species on the British List that don't breed here but are regular visitors. Perhaps the most regular is the little gull, most charming of larids. This tiny dainty thing (especially when seen alongside one of its hulking relatives) visits favourite spots on the British coast each autumn and spring, sometimes in quite sizeable flocks. It has also had a go at breeding in Britain, most recently in 2016 when a pair set up home at RSPB Loch of Strathbeg in Scotland. This pair made history by successfully fledging two chicks – all previous known breeding attempts had failed. Sadly, there was no repeat performance in 2017, but perhaps the little gull will join the list of Britain's regularly breeding gulls in years to come. If it does, it follows in the footsteps of the Mediterranean and yellow-legged gulls, both of which only began to breed in Britain late in the 20th century, having spread our way from southern and eastern Europe.

Another regular visitor from the east is the Caspian gull, a central Asian and eastern European species which is spreading westwards, and being found more and more frequently among gull flocks on the south and east coasts of England. This gull looks very like the good old herring gull that we all know well, so some pretty serious birding skills are needed to identify it with confidence. Older field guides to European birds don't even acknowledge it exists, because it used to be classed as a subspecies of the herring gull or the yellow-legged gull. Its genetics (as well as its looks and manner) support it being a real species, and when you see one side by side with herring gulls there are a few things that stand out – its little eyes, long wings, long bill and, unfortunately, rather baggy-looking belly.

Two big gulls from the Arctic – the Iceland gull and glaucous gull – appear every winter in varying numbers. The further north you are, the more likely it is you'll meet up with one or both of these gulls, which are known collectively as 'white-wingers' because the adults lack any black markings in the wings, instead sporting strikingly pure white wingtips. The Iceland is an elegant, long-winged gull the size of a small herring gull, while the glaucous, or 'glonk' as it's affectionately known among birders, is a barrel-shaped whopper that's only just outsized by the world's largest gull, the great black-back. Loiter in a harbour in northern Scotland or Ireland in winter and you stand a good chance of meeting both an Iceland and a glonk, perhaps several of each. We also play host to the Canadian subspecies of Iceland gull, the Kumlien's gull. This bird isn't quite as frosty white as the Greenland birds are, but has some shadowy grey in the tips of its flight feathers.

The beautiful Sabine's gull is much coveted by a particularly hardy crop of birdwatchers who can be found seawatching from south-western headlands during westerly autumn gales. It breeds in the high Arctic, and those that nest in Greenland and eastern Canada migrate across the Atlantic down past the equator to winter off the western coast of southern Africa. A proper pelagic bird, it is quite happy to travel well out at sea but some (especially youngsters on their first migration) lose their nerve in violently windy weather and come closer to shore. Severe storms often blow a few 'Sabs' into river mouths and further inland, where they find a quiet reservoir to recuperate before resuming their journey. However, the majority of the couple of dozen Sabine's gulls seen 'in' Britain most years are technically seen *from* Britain. This small, long-winged, slightly fork-tailed and graceful gull is distinctive on the wing, with a unique pattern of black, white and grey triangular wedges on the uppersides of its wings. The breeding-plumaged adult also sports a dusky grey hood, outlined in black, and a bright yellow tip to its dark bill.

The Sabine's gull, little gull, Caspian gull, Iceland gull and glaucous gull are all regular visitors to Britain. Being in or around Britain is not 'wrong' for them, and at the right time of year there are probably several around at various locations, but none of them are very common. Even if you go out birding a lot you probably won't see them very often, and for a bog-standard birder like me, seeing any of them is a surprise and a treat.

The next tier up are the real rarities – the lost souls, which 'should' be far away but have taken a wrong turn somewhere. They come here from North America, eastern Asia, southern Europe and the high Arctic. These are the ones that really get the blood pumping. The rarer the gull, the more excitement it sparks among birders, especially twitchers – that special breed of rarity-chasing birder whose *raison d'être* is to amass as long a list of species as possible. A really rare gull will trigger a full-scale 'twitch' to its location, with crowds of scope-bearing, camera-touting, camo-sporting humans hurrying to clap eyes on the lost bird before it flies, or dies.

Two of the most adored gulls on the British List are Arctic breeders. One of these is the ivory gull, a snow-white beauty when adult, and just as pretty as a delicately black-flecked juvenile. Often when an ivory gull turns up in Britain it stays a little while, and spends its time camped up next to the washed-up carcass of a porpoise or whale, nibbling away at the blubber until the corpse becomes, presumably, too 'matured' to be appetising. The other is the Ross's gull, a tiny and dainty example of gull-kind with a wedge-shaped tail, a sweetly wide-eyed and small-billed face, and (in breeding plumage) a lovely rosy flush to its white plumage. Being of Arctic origin, they are both likeliest to show up in the far north. The ivory gull is as closely linked to the shifting sea ice in the Arctic circle as the polar bear is (and indeed it often 'tidies up' after a polar bear has killed something), but at least it can easily escape further south to a Scottish beach if it needs to.

Gulls from further south or south-east also feature on our national list. One of them, the Audouin's gull, used to be a serious rarity in world terms, its numbers falling to only about 1,000 pairs in the 1960s, all living around the Mediterranean. A combination of conservation effort and the gull's own gullish ability to make the best of things has seen numbers climb to 10,000 pairs today. It has also expanded its range and started to turn up well away from its homelands. In May 2003, one was seen in the Netherlands. This bird decided to make two nations' worth of twitchers very happy and moved on to England a few days later, crossing the sea to make a short stay at Dungeness in Kent. A couple more Audouin's gulls have since reached Britain. This gull is distinctive, looking like a mini herring gull but with a red bill and legs, and a wash of cloudy grey over its body plumage.

The slender-billed gull is another species that's been on the increase. It breeds around the Med and Black Sea, and nearby areas, and there have been a few records of it in Britain over the years, mostly of adults which may have been on the hunt for a new breeding area – if so, they have (to date) decided not to bother to breed here. This gull looks like a lankier, spikier version of the black-headed gull (or what a black-headed gull looks like in winter, at least, because the slender-billed lacks a black hood). It often has a very definite pink tint to its plumage when breeding.

The Pallas's gull or great black-headed gull is one of the few gull species that almost every British birder, larophile or not, would love to see. I just went to take a look at its Wikipedia page and the photo at the top was a case in point – what a bird. The picture shows an adult in breeding plumage with a full, charcoal-black hood and sparkly white eye-ring, standing tall with its head tipped back and multicoloured bill wide open. This is a huge bolshy bird, only just smaller than the great black-back – it's by far the biggest of all the hooded gulls. It breeds inland, around lakes and marshes in Russia, China and Mongolia. There used to be several British records, none more recent than the 1960s, and most from Victorian times. A close look at all the old records by the British Birds Rarities Committee and the British Ornithologists' Union Records Committee resulted in all but one being retrospectively rejected, as the presented descriptions were not convincing enough. Today the sole accepted British record of a Pallas's Gull was the one found in 1859 on the

River Exe. The finder, a fisherman named William Pine, was on his own and without a camera – circumstances that would make it very hard to get any sighting accepted by a rarities committee today – but he did have a gun. As Victorian ornithologists used to say, 'what's hit is history, what's missed is mystery'. The mounted remains of this unfortunate Pallas's gull are kept in the Royal Albert Memorial Museum in Exeter.

Of the gulls that reach us from North America, the ring-billed gull is the most frequent. This gull looks like a halfway house between a herring gull and a common gull – it's medium-sized, silvery-backed and pale-eyed, with a crisply defined black ring around its yellow bill. The first one ever seen in Britain turned up in 1973, but since then more and more have headed here across the Atlantic. Often they appear in autumn and stick around for the whole winter. Several individuals have turned up at the same spots year after year, earning themselves local fan followings and affectionate nicknames. Some linger into spring and try their luck at seducing the local common gulls. Could this species, increasing fast on its home turf, become a true expat and spread to Britain as a breeding bird? Black-headed gulls have done the feat in reverse, breeding in eastern Canada from the 1970s and now firmly established there, so why not?

Some other gulls that breed in eastern North America show up on a more or less annual basis, such as the laughing gull and Bonaparte's gull. These two are both 'hooded gulls', having dark heads in adult breeding plumage. The laughing gull is a distinctive, gangly creature with a mournful expression, belying its voice and name, while the Bonaparte's is very like our own black-headed gull, but slightly smaller, with a few minor differences in pattern and colour. Other American visitors are rarer – they include the American herring gull (which is also very tricky to identify), and Franklin's gull, which is a petite relative of the laughing gull, and also the glaucous-winged gull, which only joined the British List in 2009. The individual bird in question was actually seen in 2006, but it takes a while for a sighting of a species new to Britain to be verified – even in this case, where not only were there clear photographs to support the ID, but the bird was actually trapped and a couple of its feathers collected and their DNA analysed. The glaucous-winged gull, though migratory, doesn't typically occur on the eastern side of North America, so it's quite surprising that it reached Britain. It is a large, herring gull-like species with grey, rather than black wingtips.

Others on our list are similarly surprising, perhaps none more so than the slaty-backed gull, which is not particularly migratory as a species. Who knows what circumstances led to one adult bird heading at least 4,300 miles east (or 4,650 miles west) from its breeding grounds in eastern Russia, to pop up on a rubbish dump in Essex? And the gull-finders are not finished yet. There are other species hotly tipped to reach Britain one of these days, some of them considerably more likely candidates than that wildly adventurous slaty-backed gull. What about some other North American roamers? If the glaucous-winged gull can make it here from the north-western side of that continent, why not some other western US species, like the California gull, or yellow-footed gull, or indeed the western gull? In 2016, the celebrated birder and author Killian Mullarney found the very first Vega gull for Europe while sifting through a giant gull flock on the sea shore at Duncannon, County Wexford.

Discoveries like these have sparked gull enthusiasts throughout Great Britain to scrutinise their local flocks with even more vigour. The fact that a mega-rare gull could turn up almost anywhere in town or country is rather inspiring, even to a very non-twitchy birder like me. Every birder would love to find and correctly identify a 'rare', and finding and identifying a British 'first' guarantees fame throughout the birding community. If that special bird happens to be something that's particularly challenging to identify, which is the case with most of the larger gull species, the kudos level will go through the roof.

But you don't ever need to find a rarity or a scarcity to find yourself falling in love with gulls, nor do you need to build weapons-grade expertise in gull identification. The abundant, easily observed and easily identified species that share our islands are so endowed with charm, and each in its own particular way, that they'll keep you amused indefinitely if you let them.

CHAPTER 2

Natural and Unnatural History

Great black-backed gulls

I WAS ONCE mooching along Hastings seafront, at the eastern end where the fishing net huts loom over a few little fishmongers selling the catch of the day. I paused to watch a dapper adult herring gull, which was eyeing the fish arrayed on their banks of ice outside one shop front. It actually seemed to glance around before rushing forward and jumping up to grab what I'm pretty sure was a lemon sole. The fish shop proprietor also rushed forward, with a cry of dismay, but he was too slow – the gull was already lifting skywards, the big, expensive fish hanging from its bill. I felt bad for the fishmonger, but the moment seemed to me to be quite poetic. Evolution shaped gulls into fish eaters, but their own cleverness has made so much more of them – including crafting them into thieves.

Certain people who aren't that keen on nature often ask a particularly irritating question of those of us who are. The question is 'What is its purpose?' Or sometimes it's 'What is it good for?' Or 'What's the point of it?' The 'it' in question is a kind of organism that they don't like. Most often, it's a wasp. What the question *really* means, of course, is 'How does this organism benefit me personally?' As it happens, that's a particularly easy one to answer in the case of wasps. They help pollinate our crops and garden flowers, and they murder caterpillars and lots of other insects that might otherwise damage said crops and flowers, so they help us to have food and enjoy beauty (and thus stay alive and happy). But giving a sensible answer does rather skirt around the fact that the question is a terrible one. Everything has a purpose and a point and is good for something within its own ecosystem, even if it doesn't appear to directly benefit us humans. Gulls, widely derided by humans as pointless and problematic, are no exception. Evolution shaped them to function and prosper in particular niches, and continues to do so as their niches change with the changing world.

Gulls have always reminded me of crows. Or perhaps I should say that crows have always reminded me of gulls, as it's gulls that I knew first and know best. They are both rather unspecialised in appearance, though crows in a 'living-on-land' way and gulls in a 'living-around-water' way. As with all animals, everything about them both is fundamentally built around the food that they eat. In the economics of evolution, food is the currency, and what an animal eats is reflected in every detail of its anatomy, external and internal. These are the tools it has evolved to obtain and process its food. Availability of that food dictates where in the world the animal can live, where and when it will breed, whether it spends its time alone or in groups with others of its kind, how its waking and sleeping cycles are structured, and how it competes with and otherwise interacts with the other living things that share its ecosystem.

Crows are (by and large) opportunists, able to turn their hand, or bill, to many different ways of making a living – some of them rather unsavoury by our standards. We hate to see a magpie raid a blackbird's nest, and we're not impressed by a raven hatcheting a baby bunny to death with its ridiculously huge bill (it's quite mind-boggling that these huge, powerful beasts are, technically, 'songbirds'). Gulls are similarly opportunistic, and similarly disliked by us – perhaps more for their public order offences than for murder but they, too, will prey eagerly on helpless nestlings and hapless baby mammals, given half a chance. They may not have hooked bills and ripping talons, and their kills are often, distressingly, not what you might call 'clean', but they are alert, quick, fearless and strong, and smaller and weaker animals rightly treat them with caution. They will also attack larger animals if they are incapacitated in some way.

Gulls, like crows, are also quite inventive in how they go about catching and killing prey. You might have seen them vigorously disco dancing on playing fields – that foot-pattering is done to trick worms into thinking it's raining heavily. Not wanting to drown in their tunnels, the worms move to the surface to escape the apparent deluge, only to fall prey to the dancing gulls. There are also records of herring gulls using bits of bread to tempt pond fish into grabbing range. I've often seen the Hastings herring gulls dropping mussels, whelks and other molluscs from mid-air onto rocks to break their shells, and they will do the same thing with baby rabbits and other non-flying prey. If there's water handy they will also drown non-swimming prey (those poor baby rabbits again – and, as mentioned in the previous chapter, there is a particular lesser black-backed gull living in Hyde Park which has learned to catch and drown feral pigeons, before ripping up and eating the corpses. It often performs this feat in front of crowds of horrified tourists and has done very little to improve the public perception that gulls are in any way pleasant.

Gulls are not built for this sort of predation. They have big mouths and can swallow a lot of prey whole, but even a great black-back couldn't down a whole pigeon in one. Proper birds of prey almost invariably grab their quarry with their feet, and possess curved talons to grip, and shortish but sharply-hooked bills to rip. Gulls have webbed feet and a more or less straight bill, making the task of subduing, killing and eating a large, vertebrate animal quite a challenge, but what they lack in finesse and equipment, they make up for in enthusiasm.

Predation is just a small part of the story, though, for most gulls and most crows. More important is their ability to forage and to scavenge, and make a meal out of virtually anything organic. This is how they manage, so far, to survive in a world where wilderness is shrinking rapidly to make way for the wasteland of human habitation. Humans are wasteful and that means lots of feeding opportunities for scavengers. From the underground animals exposed in soil turned upside down by the plough, to dead or dying bycatch fish dumped in the sea, wherever humans are taking what they want from the world, there are gulls and crows on hand to clean up the leftovers. When we feed the birds in our gardens, we might not intend to attract gulls but they will come, if there's food within reach. At the park, they're on hand to intercept a flying bread crust, and I've seen one land on a Hastings windowsill and batter away at the glass until a householder opened the window and handed it some food.

There are also times when they take things that really weren't there for the taking – like that lemon sole from the fishmonger's – and then there's the viral video of the herring gull sneaking into an Aberdeen corner shop and nicking a bag of crisps. We resent them for this kind of carry-on, but maybe we should admire their enterprising spirit instead – the places we see as 'ours' leave little enough space for wildlife. Nor do gulls always pick humans as the victims for their thievery. You will often see one gull trying to appropriate a meal from another, and they will also harass other birds that are feeding, in the hope of stealing a bite at least. Gulls will loiter around white-tailed eagles (and, across the pond, bald eagles) that are fishing – a risky move, as both eagles are more than capable of making a meal of the gull itself. On the Farne Islands, I've watched herring and black-headed gulls trying to ambush puffins as the latter return from sea with a bill-full of sand-eels. They are not as skilled at, nor dedicated to, piratical pursuits as their relatives the skuas, but they will never pass up an easy mark.

The question of how clever gulls really are, relative to other birds and other animals, is a tricky one to resolve. Looking at brain size (absolute and relative to body size) and brain structure is the first step in hazarding a guess. When I was a child, I found the skull of a small bird of prey while out on some suburban adventure, and brought it home, where I identified it as that of a kestrel – back then, a rather commoner bird than it is today and very much the 'default falcon' in most habitats. Some months later I found a herring gull skull down on the beach, and I decided to go in for some amateur science and compare the two birds' cranial capacities – the volume of their brains, in other words. So I poured some budgie seed into each skull's brain case 'til they were full, then decanted and measured the amount. Much to my surprise, they were the same. I'd been sure the gull skull would have a bigger brain case. It's a much bigger bird, for a start – nearly twice as long, bill-to-tail, and at least three times as heavy as the falcon. But the surprise was mainly because I think of gulls as super clever.

It also turns out that the forebrain/brain stem size ratio, a pretty good indicator of intelligence, is quite small in gulls (to simplify hugely, the forebrain can be thought of as the 'thinky bit' and the brain stem as the 'instincty bit'). A 2012 study by Z A Zorina and T Obozova on avian intelligence looked at this ratio in various bird groups. The researchers noted that their chosen gull species, the glaucous-winged gull, had a forebrain/brain stem ratio of about 4:1. In crows, it's about 16:1, and even little birds like finches and tits are faring much better than gulls with a ratio of about 8:1. In terms of brain size and structure, things don't look too promising for gulls in the avian mastermind league.

The meat of the study concerned testing the birds' cognitive abilities. The researchers worked with wild glaucous-winged gulls on their breeding colony, and presented the birds with a series of mental challenges, most of which they failed miserably. They didn't realise that a hidden object still existed (or at least didn't search for it). If they were shown a food treat being slowly moved behind a screen, they did not work out how to relocate it. They could learn to tell a big box from a small box, but could not tell a group of two same-sized boxes from a group of five. These tests make gulls about as clever as pigeons, and a lot dimmer than crows, which are easily successful at the same kinds of tests.

Most early research on avian intelligence was done on pigeons, and extrapolated across all of bird-kind, leading to the still widely held view that birds are a lot less clever than most mammals. However, once research focus broadened to include the likes of crows and parrots, it became clear that the spectrum of bird braininess was a broad one. Crows do better than cats, for example, on these sorts of tests, and better than many primates on some others. Some scientists consider that ravens and larger crow species are as clever as the great apes (ourselves not included, presumably...). They're capable of advanced tool use (using a tool to make another tool), can recognise themselves in mirrors, and recognise and remember different human faces. They can also do abstract reasoning, count, and reason about the intentions of other crows – for example, if they hide some food and then realise another crow is watching, they'll move it to another hiding place.

I admit I was disappointed to learn that gulls are, by these measures at least, quite the bird-brained thickos. I was surprised, too, because gulls always seemed to me to be pretty good at the sorts of innovative and exploratory behaviour that looks like intelligence to us. That crisp-stealing herring gull mentioned earlier – at some point it had to work out that the shop contained crisps, which were themselves contained in bags, which needed to be held down and torn open. That's a complex series of behaviours right there, which surely takes a bit of brain power to string together. I've since seen a couple more videos of gulls stealing bags of crisps from shops – in one case the gull had to negotiate an automatic sliding door. Looking for these on YouTube led me to an amusing 'birds stealing food compilation' video, in which five out of nine of the avian thieves were gulls. One actually flew in through a partly open window to pillage the full plates on a kitchen worktop (though it did then bump into the not-open bit of the window when trying to leave).

Further investigation did find some support for my notion that gulls are quite clever. The ornithologist Louis Lefebvre carried out a review of published literature documenting incidences of 'novel feeding methods' observed in wild birds, and he found that gulls were high scorers, right up there with the clever corvids. Maybe gulls don't ace the formal IQ tests, but when it comes to working out how to get an easy meal, they are ingenious indeed.

Most gulls will eat most kinds of foods, but the smaller, marsh-nesting gulls have quite an unusual diet by gull standards – the little gull in particular is almost entirely insectivorous during the breeding season. I've never had the pleasure of watching little gulls on their breeding grounds, but I've seen them on migration, flying over lakes and lagoons, and their manner of flight and foraging recalls a swallow or a martin. They are, ecologically, closer to some of the tern species than they are to other gulls – at least in summer. But in winter, they become more typically gull-like in their habits. The melee of gulls that gathers at 'The Patch' at Dungeness – a warm-water outflow from the power station into the sea, which attracts lots of fish – often includes a little gull or two in autumn and winter.

Most gulls eat quite a lot of fish, but supplement this with other animal prey, as well as with carrion and, in some areas, food waste left by humans, of both animal and vegetable origin. Dropped chips don't hang around very long in Hastings. If you stroll along the Hastings seafront early on a Sunday morning, you stand quite a good chance of seeing herring gulls picking at actual human vomit, too. That covers nearly every kind of organic matter that there is. The main omission is the waste which other animals excrete from the other end … and even that may be worth eating, if you're a hungry gull with low standards. Gulls that live near seal and sea lion pupping beaches clean up placentas as well as the corpses of any mother or pup seals that didn't make it, but sometimes their interest goes beyond that. The kelp gulls on Guafo island, Chile, have developed quite a taste for baby seal poo, but only when the poo is full of hookworm larvae (a common parasite of the fur seals in this area). Such is their enthusiasm for this fare that a baby fur seal only needs to look slightly in need of a poo to attract the gulls, and their over-eager attentions leave many baby seals with nasty perineal injuries.

Having a keen eye for even the most bizarre feeding opportunity means that gulls, or at least our more familiar British species, are not really tied tight to particular habitats, and are certainly not restricted to the coast and sea. Some gull species have always nested inland. As I write this, it's early spring and I am in north Scotland, a good 30 miles inland and surrounded by Caledonian pine forest. Yet black-headed gulls are a constant presence in the skies overhead. Yesterday I found a marshy little lochan within the forest and at this tiny lake there were three pairs of black-headed gulls, taking it in turns to wheel in high circles over the water before settling on the soggy sphagnum mats or bathing in the open water. They are, I feel sure, getting ready to nest there, and I see no reason why they should not succeed.

Back home in Sevenoaks, also well away from the sea, things have been starting to sound a lot like Hastings this spring, with a few pairs of herring gulls apparently prospecting for nest sites on rooftops in my road. Several times I've heard them giving the aptly named 'long call' that signifies territorial activity, to each other – it sounds like they want to breed here, though whether they will be allowed to do so is another matter.

Not all gulls are happy in a variety of habitat types, though. Some have quite exacting habitat needs. Among our own species, the kittiwake is the one that comes to mind as an outlier, in that (curiously enough) it's the only one that's a true seabird, happy to live offshore for months on end, and nesting within the sound of crashing waves. It'll be a strange day indeed when kittiwakes start heading inland to colonise Home Counties' rooftops or forest ponds. There *is* a famous colony on the Tyne Bridge on the Newcastle/ Gateshead quayside, in north-east England. Nearly 13km from the sea, this colony is nevertheless connected to open sea by the tidal Tyne, and the kittiwakes find the bridge's narrow sills a perfectly good alternative to sea-cliff ledges. These birds were in fact originally nesting on the nearby Baltic Flour Mill in Gateshead up until that building underwent extensive renovation in the late 1990s – it's now an art gallery and a few pairs of kittiwakes have returned there.

Further afield, the lava gull of the Galapagos islands is unusual in that it doesn't like to nest anywhere near others of its kind. Perhaps all gulls would do the same if they lived somewhere with so few predators to worry about. Its Galapagos neighbour, the swallow-tailed gull, is an oddity in another way – it is more or less nocturnal. During the breeding season, both adults stay at the nest in the daytime, and one heads off at dusk to spend the night hunting at sea. They are particularly active and successful on the darkest nights with the newest moon, as this is when the little fish and squid come up from the depths – like other gulls, swallow-tails are not much for deep diving and take most of their prey near the surface. Their responsiveness to the lunar cycle replaces the awareness of the day-night cycle that most birds have.

Birds in general rely most heavily on two senses – vision and hearing. Like most other birds, gulls have acute eyesight and superior colour vision to our own. We have four types of photoreceptor cells (ones that react to light) in our retinas – rods, which handle light intensity, and three types of cones which react selectively to different colours of light. In birds, there are six. They have rods, and not three but four types of colour-sensitive cones, as well as 'double cones' which appear to be important for motion detection. Most birds have at least twice as many photoreceptor cells per square millimetre of retina as humans do. For birds, a broader swathe of the electromagnetic spectrum presents itself as visible light – they can see ultraviolet light, which we cannot. Gulls and other seabirds also have specialised red-pigmented oil droplets in their cone cells, which help them to see through the water when flying low over the surface or swimming.

Their hearing is also quite acute, though probably a bit less than our own, and they are also sensitive to a narrower range of frequencies than we are (a range of 100Hz to 3,000Hz is typical for a gull, compared to 20Hz to 20,000Hz for the average young human). Gulls' own range of distinct vocalisations suggests they are good at discriminating one sound type from another, and they are certainly highly reactive to loud noises – like mammals, birds can experience pain and sometimes even damage from noises that are too loud. In my childhood in Hastings, the beach-launched lifeboat service would summon its crew of local volunteers (and warn the fishermen of an imminent launch, as well as hopefully reassuring the actual vessel in need of rescue) by firing off a couple of 'maroon' rockets. These exploded with a pretty tremendous bang, and the gulls' reaction was instant. It seemed like every herring gull in town would freak out when the maroons went off, lifting up from its rooftop and calling at the top of its voice. So in the wake of the bangs there was a cacophonous racket of screaming gulls and, if it were daylight, you would see them all filling the air, wheeling above the town in noisy distress for several minutes before gradually beginning to settle down. Studies show that the sound-detecting cells in birds' inner ears do have the power to regenerate after damage, so they have an advantage over us there – for people, loudness-induced damage is usually irreversible.

For gulls, smell is also quite an important sense. Olfaction (to give it its scientific name) is not especially well-developed in most bird species but some seabirds excel at sniffing out food. Shearwaters and petrels use odours not only to find tasty, rotting whale carcasses but also to navigate over vast tracts of ocean, and even racing pigeons use their sense of smell to help find their way home from an unfamiliar release site. Satellite-tracked lesser black-backed gulls from Finland and Russia have been shown to mess up their migration when their sense of smell is temporarily blocked, failing to stop at their usual destination of Lake Victoria in eastern Africa. Another group of gulls in the study were nasally unimpeded, but were captured and re-released at a point outside their usual narrow migratory path. These birds still found their way back to the lake without a problem. The researchers behind this 2015 study, a team from the Institute of Avian Research in Wilhelmshaven, Max Planck Institute of Ornithology in Radolfzell and the University of Konstanz, suggested the gulls follow a series of 'smell-posts' as they navigate, detecting the distinctive aromas of the Black Sea and Nile Delta from their flight path, from distances as great as 300km.

Birds don't usually have well-developed taste buds (that would explain why gulls happily eat the most horrible detritus). The sense of touch is another matter, and there is some evidence that gulls that often forage at night, such as the grey gull of South America, *Leucophaeus modestus*, use tactile tricks to find food, rather than having specialised eyesight adaptations for operating in low light. When you see gulls 'tap-dancing' on soft grassy ground to lure worms to the surface, it seems feasible that they can sense subterranean movement as their quarry approaches, but as far as I can see, no one has actually studied this yet.

Regardless of species, there are several elements of the gull lifestyle that are almost universal, the most obvious of which is that gulls are social. One of the reasons that gulls are so appealing to that particular subset of birdwatchers, the larophiles, is that outside of the breeding season they form big, mixed-species flocks wherever conditions are good for feeding, resting or sleeping. You don't have to hunt around for one elusive little bird – you can settle down and work your way through a large collection of big, easily observed birds, which are individually variable enough to satisfy anyone who really enjoys a bird identification challenge. These big mixed flocks also offer the chance to watch social interactions, between different species and members of the same species.

In the breeding season, gulls head off to their breeding grounds but they stay sociable. Often a colony of one species will have a few pairs of another, and some species do form truly mixed colonies. However, when you watch gulls going about their business you might soon get the impression that these birds actually can't stand each other – low-level violence and screaming matches are commonplace in both breeding and non-breeding gatherings, and can break out with minimal provocation.

One of my favourite places to take bird photos is in Kensington Gardens in central London. There are a few hidden gems in this groomed green space. Walk through a particular avenue of trees and ring-necked parakeets – those rowdy green non-natives now thriving so impressively in our capital – will fly down to check your hands and head for hidden stashes of food. Stand at the foot of a particular sweet chestnut tree, in just the right spot, point your binoculars towards the crown of the tree and you will (if you're lucky) meet the furious yellow stare of a male little owl, guarding his mate in her nest hole deeper within the tree. It's also a wonderful place to photograph gulls in winter, with lots of black-headed, common, herring and lesser black-backed gulls on the Long Water and Round Pond. When not harassing park-goers for bits of sandwich, some of the gulls rest on the rows of low posts that punctuate the Long Water here and there, and make for pleasing images – a mixture of species, lined up neatly, all facing the same way. But watch for a while and sooner or later a newcomer will arrive. Watch that bird. It won't settle on a spare post. Instead, it will, almost invariably, go for one that's occupied. The gull that's already there will either jump off with a squawk of annoyance, or it'll stand its ground – only for the new arrival to land on its back and stay there until the first gull tires of being stood on and tumbles off. Frequently this action will dislodge the back-sitter too and both of them will fly off, swearing at each other as they go and leaving the coveted post vacant. It all seems tremendously petty – why do they do it?

Sociality in animals is often a rather tense affair. Survival means selfishness, after all, when you are working hard for every mouthful of food, when you need to find the safest possible place to sleep, and when you want to make sure you're not the one who gets caught and killed by a predator. Successful breeding takes cooperation with a partner, but you remain in competition with all the other couples – you all want the best available nest site, and to get hold of as much food as possible for your young. Gulls put up with the competition because being sociable brings benefits too. More eyes to spot danger. More muscle, so a dangerous predator can be attacked and driven away rather than avoided. Group vigilance allows gulls to build their nests in open, exposed places that wouldn't be safe for lone individuals. Being gregarious outside of the breeding season can also hold advantages – you continue to benefit from group vigilance. Also, resting and roosting in flocks can help save energy on cold days and nights. You want to be as close to the middle of the group as possible – for safety as well as warmth – and the bigger the group, the bigger the optimal middle zone will be.

It's a bit different with feeding. When gulls converge on a big food source, they all stand to benefit (except for the one that found it first and has to share), but this is not cooperative. You can see the dynamic anywhere where gulls gather to scavenge – if any individual grabs a bit of food that's too large to be downed in one quick gulp, it will run away or fly away with a whole gang of other gulls in hot pursuit. The finder desperately tries to choke down its prize before a bigger gull piles in and steals it. The chases can last for ages, if the bird with the food can't manage to swallow it quickly, and it may decide to cut its losses and give up its meal for the pursuers to fight over. Stealing food from other non-gull birds is not necessarily that frequent, but stealing it from other gulls is part of the most fundamental larid skill-set, and predictably it is often big gulls stealing from small ones (though the smaller species are more agile in the air so stand a chance of getting away with the goods).

So, even though they don't necessarily like each other very much, all of our gulls are gregarious when they're not breeding. When it comes to nesting time, the smaller gulls are most inclined to form large, dense colonies, while larger gulls go for more flexible arrangements, depending on the terrain. The herring gulls nesting on Hastings rooftops are quite spaced out, because suitable spots for a nest are limited, but those that nest on the ground on Havergate Island in Suffolk are more closely packed (and share the limited space with an even higher number of lesser black-backs).

My nearest black-headed gull colony is the very big gathering at Rye Harbour, in East Sussex. Here there are several old gravel extraction lakes, and their stony shores and islands offer space for plenty of gulls to nest – some 1,400 pairs of black-headeds. There are also terns nesting here – Sandwich terns with their rakish shaggy crests, and sleek, piercing-voiced common terns. The terns are peerlessly swift and graceful on the wing and do a great job of mobbing any passing bird of prey, but on the ground they are awkward, with their short legs, and the bigger, faster-running black-headed gulls have no difficulty pushing their small cousins to the outer edges of the colony. However, the black-headeds must themselves concede some ground to their own bigger, badder relative, the Mediterranean gull. I look to the centres of the islands to find the Meds. There they stand in pride of place, the biggest birds in the colony – at least a couple of dozen pairs in most years. I need to look to the outskirts of the groups, close to the water, for the terns. Every other scrap of shingle is taken up by black-headed gulls. Another black-headed gull colony I spent some time watching is at RSPB Leighton Moss. These gulls had the place to themselves and nested on marshy mats of vegetation on lake islands, among clumps of pretty yellow flag irises.

In early spring, the adult black-headed gulls show up on their breeding colonies, already mostly, if not fully, dark-headed. Being first on site means a good chance of securing a prime position, but you run the risk of going hungry, because most good nesting areas are not very food-rich until later in the year. Those that stay a bit longer at well-provisioned wintering sites may be in better condition and thus able to push the less well-nourished early arrivers out of the top positions. This juggling act dealt with, the gulls need to find their partners. Like other gulls, they are monogamous, even though they go their separate ways in winter. In an ideal world, a black-headed gull that bred successfully last year will find the same mate again, and they'll nest in the same spot again (if not a slightly better spot). However, those that failed last year will try for a different spot and, perhaps, a different mate.

In courtship, it is usually easy to tell the female from her partner – as with other gulls, the males are bigger and heavier. The pair strut in circles around one another in a very tall posture, with the wings slightly open, and wag their heads from side to side ('head-flagging'). Periodically they both bow down low with the neck stretched forwards and head tilted up, and break into a run. The object seems to be to show off their hoods and white crescent eyelids from every possible angle. If the female is lucky, after a little strutting the male may then vomit up some food for her. A version of this display happens whenever the pair is reunited, and any neighbour that wanders into their strutting path will be chased away. Even in a dense colony, each pair vigorously defends a small circle of space around their own nest.

The nests themselves are nothing to write home about really – like most gull nests they're scruffy heaps of whatever vegetation the birds can find to hand. Or maybe there is artistry in them that escapes our human sensibilities. I once watched a gull making a tremendous meal of adding a new and much-too-large stick to its nest, trying it this way and that, while its mate clearly had quite different ideas and kept trying to grab the other end of the stick.

Once there are eggs in the nest, the gulls' paranoia is ramped up to a new level. In this, they are like some of their non-gull relatives – the terns, and also certain waders such as avocets. Like black-headed gulls, these birds prefer to nest in colonies. Anyone who has made the birders' summertime pilgrimage to the Farne Islands off Northumberland's delectable wild coast will have experienced the wrath of Arctic terns in full ferocious attack mode. On Inner Farne, they nest in profusion, and many choose to place their nests just a footstep off the main path, as well as in the little courtyard by the loos. These areas are good nesting spots because the regular presence of people discourages the most dangerous nest predators – gulls. However, the terns have not quite put two and two together and feel very threatened by the conga line of hat-wearing birders who make their way from the landing point up to the crown of the island, past umpteen tern nests. Those sitting on eggs alarm-call constantly, a bone-hard rattle that rings in your ears hours after you have sailed back to the mainland. The hat you wore (if you were wise) protects you a little from the dagger bills of the sitting birds' partners, which circle, dive and stab relentlessly until you are well beyond the danger zone. One fierce tern might not be too troublesome – it's still quite a little bird after all, but it's easy to see how even a determined predator would lose its zeal in the face of an onslaught from hundreds.

With their long, spindly legs and needle-slim, comically upturned bills, avocets are more box kite than fighter jet. These dapper black-and-white waders look barely able to defend themselves, let alone their nests, but what they lack in fighter physique they make up for in fighting spirit. I was once watching a pair of avocets as they carefully shepherded their four tiny new chicks on a small, muddy island, close to the shore of a large, shallow flood. They were alone – no other nests anywhere nearby. A little egret came winging along, heading towards me from the far edge of the water to the fields beyond. This dainty, small heron looked delightful in its pristine-white plumage, so I turned my camera away from the baby avocets to photograph it. Its flight path was taking it near the avocets' island, and the avocets' response was instant and decisive. One bird rose up at once and flew at the egret, 'pleeting' furiously, and aimed a kick. Caught squarely on the top of its head, the egret crashed and sprawled into the shallow, muddy water, just as the second avocet flew up to join in. My photos record a chaos of wings and legs and great gouts of brown water flying skywards as the egret managed to right itself and stagger clear enough to take off again. It left the area in great haste, no longer pristine white but besmirched with mud and croaking in alarm as it went. I don't think it so much as cast an eye at the tiny avocet chicks, but egrets (like other herons) are predators and the parent avocets were not taking any chances.

Avocets, Arctic terns and black-headed gulls are evolutionary cousins, evolved to nest on the ground. Choosing islands helps protect their nests from mammalian predators, but against winged hunters, attack is their only defence, and they all use coordinated vigilance to defend their nests and young. A bird of prey passing high overhead may only draw a few of the colony up in response, but any predator that actually targets the colony had better prepare itself for a robust response. Even on the mainland, a vast enough colony can team up to see off a fox or a stoat.

I saw the Rye Harbour black-headed gulls in action against a predator one day in early summer and the sight was impressive – the cloud of gulls that rose up from the island as one was several hundred strong and so dense that I couldn't pick out the target of their screaming fury. My camera, however, did. Examining the photos of the incident later on I saw, dead centre of that maelstrom of wings, an adult lesser black-backed gull lifting upwards, and all the efforts of those black-headed gulls had been for nothing, because from the lesser black-back's bill dangled a lifeless black-headed gull chick. There is no way to eliminate entirely the risk of nest thievery.

And yes, gulls kill other gulls, and that's a little shocking, perhaps. They will even sometimes kill chicks of their own species, within their own colony. But the crux of the matter is that a large colony of hypervigilant gulls, acting as one, are still not necessarily able to defend their young from predators. Make the colony a bit smaller, and their odds of success shrink accordingly. Neighbours might be tremendously irritating and occasionally dangerous, but they have their uses.

Kittiwakes don't nest on flat ground but on precipitous ledges, and this in itself grants them a high degree of protection from predators – no mammals can get anywhere near their nests, and even aerial predators will find it tricky. Cliff-nesting does, though, introduce a new enemy – gravity. The chicks of gulls that nest on the flat are very active from a young age and wander about, dodging attacks from neighbouring adult gulls and straying further from the nest as they get older and stronger. For baby kittiwakes, it's not an option to wander very far, because the ledges are narrow and they will very soon reach either a neighbour's nest (whether another kittiwake's, or the nest of a guillemot, razorbill, shag, fulmar or other cliff-nesting species) with no way past, or they'll find themselves teetering over a deadly drop. You will thus see well-grown young kittiwakes sitting down placidly in their nests right up until they are ready to fledge. This behaviour also helps keep them safe from predators such as great skuas and great black-backed gulls, which might fly close along the cliff-face and try to grab an unwary baby bird on the way.

Being seafaring gulls, kittiwakes gather most of their nesting material from the sea and shore, and in mid-spring you'll see them returning to their cliffs carrying clumps of seaweed, which they plucked from the strandline or found floating on the sea. The seaweed pile is trodden down in the middle to make a deep cup, in which the eggs are laid. Unfortunately, though unsurprisingly, kittiwake nests today often include a certain amount of plastic debris as well, and because this is often strong and stretchy, rather than brittle like dry seaweed, it can't be trodden into shape very easily and can even fatally ensnare nestlings.

Larger gulls often nest alone. The great black-backed gull, which fears neither man nor beast nor any other bird, is usually a solitary nester, though may nest on the outskirts of other seabird colonies. Herring and lesser black-backed gulls are more inclined to form loose colonies, whether on undisturbed shorelines, clifftops or rooftops, but they too may nest well away from others of their kind, if they find a safe place to do so.

Let's talk a little bit more about 'others of their own kind' for a moment. Most wild birds don't naturally hybridise with other species. There are exceptions, of course. Ducks are notorious for randomly copulating with anything that stands still long enough – only yesterday I opened Facebook to be confronted with a photo of a mallard enthusiastically mounting an alarmed-looking chicken. Gulls also hybridise more than most other birds do – a consequence of living in mixed colonies, most likely. In Britain, the most frequent hybridisation is between lesser black-backed and herring gulls, and black-headed and Mediterranean gulls, because they often nest shoulder-to-shoulder. If you're a gull that's all fired up and ready to breed, but can't find a partner of your own species... well, it's easy to see how it could happen. There may be a minor language barrier to overcome during the courtship phase, but most closely-related gull species do seem to be able to cross-breed successfully.

Needless to say, identifying hybrid gulls is even more of a fun challenge than 'normal' gull ID. Further afield, hybridisation is very common in certain gull populations. Across the pond in western North America, the glaucous-winged gull occurs in the north and the western gull further south, and where their populations meet (between Washington and Oregon) they interbreed so freely that a so-called 'hybrid swarm' has developed. The hybrid gulls can look like perfect intergrades between their parent species, or much more similar to one than the other. They are fertile, so carry on having second-, third-, fourth- and so on generation hybrid young of their own. A hybrid swarm can sometimes actually take over the range of both parent species and effectively eliminate them.

Some hybrid combinations occur often enough to have earned their own special names in the birding community. For example, the hybrids between glaucous-winged and western gulls mentioned above are nicknamed 'Olympic gulls'. A herring gull × glaucous gull hybrid is a 'Nelson's gull' or a 'Viking gull', and glaucous × glaucous-winged is a 'Seward gull'. In Canada, there is a population of gulls that look intermediate between Iceland and Thayer's gulls. These birds, known as Kumlien's gulls, might be a hybrid swarm between Iceland and Thayer's, or they might be a subspecies of Iceland gull. Given that some ornithologists reckon Thayer's itself to be a subspecies of Iceland gull, maybe it's best to leave this one well alone.

Not only will gulls sometimes pair with the 'wrong' species, but they don't necessarily stick to traditional male-female pairs, either. Researchers George and Molly Hunt, studying western gulls at a breeding colony in California, found that 14% of the pairs tending nests were female-female – the giveaway was often that the clutch was larger than usual, with both birds contributing eggs. Though the eggs were usually fertile, showing that the birds did at least spend some time with males, through the incubation and chick-rearing period the paired females are devoted to one another and show the full range of courtship and bonding behaviours. Homosexual pairings have been documented in at least five other gull species.

Caring for the babies is a two-bird job, as is the case with most birds (and this is why they form pair-bonds, for the duration of the breeding season at least). Gull chicks are what's known as semi-precocial – one of four categories of development that a baby bird can be. Altricial chicks are the ones that are born naked, blind, strikingly ugly and almost completely helpless – most songbirds are like this. Semi-altricial chicks are born open-eyed and fluffy, but still pretty helpless – birds of prey tend to fall into this bracket. Fully precocial chicks hatch out covered in fluff and open-eyed, *and* they can run around and feed themselves very soon after hatching – examples include ducklings and baby pheasants, and also most baby waders. Gull chicks are just a stage behind – they can move pretty well soon after hatching but need to have food brought to them by their parents for several weeks. When they are very small, they also need to be kept warm by a parent, so one of the adults will stay at the nest at all times, while the other is away foraging.

Gulls bring food to their babies in their crops – in other words, they swallow it, fly back to the nest, and puke it up again for the babies to eat. Rather gross from our point of view, but the young gulls don't seem to mind at all, and it is much easier to fly a kilometre – or much, much further – with a full crop than with a full bill. Indeed, it's puzzling that there are some seabirds that *don't* do this. Wouldn't puffins have an easier time if they swallowed their sand-eel catch to regurgitate later, rather than carry it in their bills? The baby gulls encourage their parents to disgorge the goods by following them around, calling and pecking at the adults' bills – their begging becomes more and more vigorous as they get older.

Most gulls have at least two babies per brood and usually more, so there is competition between the youngsters at each mealtime. The youngest and smallest can be pushed aside and sometimes attacked, but as long as there is enough food around, the siblings usually have peaceful relationships with one another. That's not the case between chicks from different families though – for example, black-headed gull chicks in a colony will regularly try to beat up their neighbours.

Over the few weeks that the gull chicks spend at or near their nests, their fluffy, grey down is gradually replaced with soft, grey juvenile plumage. If space allows they will roam further and further from the nest, far enough to get into scraps with the neighbour's kids, which can be quite ferocious. If they are near water, they may start swimming well before they are able to fly, and they'll also explore their environment, investigating all manner of objects with the same curiosity that will serve them well throughout their adult lives. However, they will always race back when a parent turns up with a food delivery. As their flight feathers grow, they spend more and more time exercising their wings – jumping, flapping and eventually flying. They will still need to be fed for some weeks after they first begin to fly, but will gradually work out how to find food of their own. Parental care may continue sporadically for months, though. I have seen a young herring gull strenuously begging from what I presume was its mother or father in March, meaning it was about 10 months old and should have been taking care of itself for the last six months at least. (To be fair, the adult did not oblige and instead just stood there looking as incredulous as a gull can look, while the youngster pecked at its bill and whined.)

Once breeding is completed and the juveniles are at least mostly independent, gulls of all ages become more social, and stay that way for some months. They are moulting, and even the breeding adults have no serious responsibilities any more, so they spend a lot of time preening, bathing and just generally loafing around. Food is still relatively easy to find, so foraging need not take up more than three or four hours a day.

When I go to Dungeness in the autumn, my progress along the little coast road is frequently arrested by the need to pull over and look through a loafing flock of gulls. They stand or sit closely packed on the shingle, in groups of fifty to a hundred, all in a rather hunched-up posture and facing the same way into the breeze (because the wind that sweeps along this peninsula is nearly always strong and cold, and you don't want it getting under your feathers). They stare into the middle distance, looking enigmatic, and only bother to move if an unleashed dog or intemperate child comes hurtling their way. Here, the flocks are mostly of great black-backs and herring gulls, with clusters of black-headed gulls keeping a cautious distance away from the big boys. But there might be something else among them, and this likelihood increases the deeper we get into autumn. One time, it was a first-winter Caspian gull from eastern Europe, looking rather dainty and frosty-headed compared to the chunky, dusky young herring gulls surrounding it. Another time there was a similar-aged glaucous gull among the regulars, a beefy Arctic monster the colour of very milky coffee. When gales swept the south coast in winter 2013, there were even a few kittiwakes hunkered down on the beach here, the open sea for once too scary even for them.

What to do about winter is an issue that different gulls handle in different ways. The established pairs of rooftop herring gulls will often spend at least some time near their nest sites all through the winter, but herring gulls that nest on flat beaches won't. They move to other parts of the coastline, or head inland – anywhere where food can be found. Kittiwakes abandon their breeding cliffs wholesale and embark on several months of ocean wandering, mainly way out in the north-western Atlantic, and many of the British-breeding lesser black-backed gulls undertake a proper southward migration, to spend winter in Spain, around the western Mediterranean, and down the coast of western Africa. They might cover in excess of 4,000km on these journeys.

Conversely, some of the gulls we see in winter are visitors from abroad. Out of more than 2,000 foreign-ringed common gulls to be found or seen in Britain, 732 were from Norway, and more than 200 each from Sweden, Denmark and Germany. Britain enjoys a milder winter climate than mainland Europe, and has more coastline for its size as well (not that common gulls are strictly coastal by any means). We also receive some very significant numbers of Mediterranean gulls and black-headed gulls from mainland northern Europe. Of the Med gulls ringed abroad and found here, 2,352 of them were from Belgium, 1,121 from Poland and 1,123 from Hungary, and a few hundred each from France, Germany and the Netherlands. Black-headed gulls come to us in profusion from Norway, Denmark, Sweden, Finland, Estonia, Lithuania, Latvia, Poland, Germany, the Netherlands and Belgium, and there's some interchange between here and southern Europe and northern Africa too.

The great black-backed gull's map shows less diverse movement than the smaller species, but birds do move between the British Isles and the Atlantic coast of mainland Europe, all the way from northern Spain up to northernmost Scandinavia, and also Iceland. The kittiwake, comfortable out over the deepest oceans, is the biggest traveller of all our gulls and the only one to regularly make a full Atlantic crossing. An impressive 71 British-ringed kittiwakes have turned up in Canada, and 186 in Greenland. There are also recoveries from north-west Africa on both the Atlantic and Mediterranean coasts, as well as many northern European countries, and various random spots in the north Atlantic, hundreds of kilometres from the nearest land. The longest journey was made by a chick born on the Farne Islands in 1967, and found dead in Massachusetts in October 1983. A total of 5,198km of mostly open sea separated its start and end position, but that bird could have made the crossing many times in its 16 years of life. We can't know for sure, but it almost certainly didn't move to the USA permanently. Most kittiwakes either return to their natal colony to breed, or join another one within 100km – only about 21% of young birds settle further away than that, and virtually none set up home beyond 900km from their birthplace.

Ringing recoveries have lots to teach us about gull movements and migrations. However, it's a hit-and-miss affair. Most birds that get ringed will never be seen again, alive or dead. But now we're in a new era of technology, where we can fit birds with tiny tracking devices, and thus follow individuals' journeys remotely, no matter how far they wander. Two main kinds of trackers are used for these kinds of studies. Geolocators work by recording light intensity over time, from which the bird's position can be worked out. It's necessary to recover the geolocator to get hold of the data, so this kind of tracker works best when placed on adult birds caught at their breeding grounds – you've a good chance of being able to catch the same bird again the following year when you know where its nest is. Then there are GPS satellite trackers, which enable you to follow the bird's exact route in real time – better, but also bigger, and pricier. You may not ever get them back, either – they're designed to fall off after a few months or years.

Geolocators have been used on some gulls in recent years. A study led by Dr Maria Bogdanova from the Centre for Ecology & Hydrology (and published in 2011 in the Proceedings of the Royal Society B journal) looked at what UK-breeding kittiwakes got up to after the breeding season, and uncovered the surprising truth that those which have a failed breeding season behave quite differently to those that did rear young. The failed breeders travel some 3,000km to winter off Canada, while those that bred successfully spend winter 'at home', just a few kilometres offshore from their breeding colonies. Another study on kittiwakes, this time using GPS trackers, looked at the birds' foraging behaviour through the breeding season and found that they hunt in different areas during incubation than during chick-rearing, travelling further during the former than the latter. This study, carried out by G S Robertson and colleagues and published in *Marine Biology* in 2014, also found variation in foraging areas from year to year.

The British Trust for Ornithology carried out GPS tracking studies on lesser black-backed gulls for a few years from 2010. Some of the birds, tagged on their breeding grounds in East Anglia, migrated down to Gibraltar via northern France, then directly across the Bay of Biscay to northern Spain, hugging the coastline for the rest of their journey. On the way back in spring, though, they took a much more direct track, zooming straight up the middle of Spain. This makes sense – there's no great rush to reach the wintering grounds, but in spring you definitely want to arrive promptly in case another gull is going to try to move in on your mate, or territory, or both. Not every tagged bird went this far – some didn't even leave England and instead passed the winter on the south coast. The same birds didn't necessarily go to the same wintering sites either. Having a bit of flexibility in one's attitude to migratory routes and locations is a good thing in this fast-changing world.

If a gull makes it to adulthood, it stands a fair chance of living a few more years, especially in the case of the larger species. Predation is a major cause of death for most bird species, but very few predators could tackle a healthy adult great black-back, while many would attack a black-headed gull (including the aforementioned adult great black-back). As well as meeting its end at the bill of a bigger, badder gull, a gull of one of the smaller species might be taken out by a peregrine falcon, goshawk, buzzard or even a bold sparrowhawk, while golden and white-tailed eagles are a rare but real hazard to gulls in northern Scotland and Ireland. It could also fall victim to a mammalian predator such as a fox, stoat or even a domestic cat, none of which would be likely to tackle a great black-back. At sea, great skuas pose a risk to smaller gulls, while dolphins and even seals might opportunistically have a go at gulls of any size. Nevertheless, survival rates in all six of our widespread gull species are high – around the 90% mark (which means any adult gull has a 90% chance of still being alive one year on).

Besides eating them, being eaten by them and stealing food from them, gulls interact with other bird species in various ways. Take a look across the lush flat fields of the Romney marshes in winter and you'll see black-headed gulls rubbing shoulders with the likes of lapwings, greylag geese, wigeons, starlings and curlews. Each species has its own way of getting a meal out of the field – for example, the geese and ducks crop the grass, while the curlews probe deeply with their absurdly long bills to reach deep into the soil for worms and the like. The gulls are there in their usual opportunistic way, grabbing what they can. If a peregrine turns up, all of the ground-feeders will take to the air, creating a stunning spectacle of multi-tonal wings. Come spring and summer, most other birds become hostile towards gulls and will chase and mob them to keep them away from their nests.

While writing this chapter, I vaguely remembered hearing a story, some years ago, of a brood of abandoned herring gull chicks which were cared for by a pair of peregrine falcons. I checked online, and to my great joy, found an amazing video showing exactly this. The three fluffy, grey gull chicks were in their cliffside nest, somewhere in south-west England. A peregrine flew in clutching a feathery lump of prey, settled among the chicks and began to tear off bits of meat and carefully feed them to each baby gull in turn.

The video was posted by the South West Peregrine Group, who'd found and monitored the nest. I went off to their website to learn more, and found that this recording was actually not of the incident that I recalled, but a later one. The one I'd had in mind had occurred in 2001, where a pair of peregrines cared for two herring gull chicks for a short while, but then the chicks were attacked and killed by adult herring gulls. The video I'd watched was made in 2014, and in this case all three of the baby gulls are thought to have fledged successfully.

Cross-species fostering is a thing that happens from time to time in nature. There are, of course, some species that do it on purpose and make their living this way – the cuckoo being the most familiar example in Britain. Accidental cross-fostering like this is another matter though. It happens now and then with closely-related species nesting in mixed colonies – it's been noted in gull colonies, for example, with herring gulls rearing baby lesser black-backed gulls and vice versa. But gulls and falcons are vastly different in behaviour and ecology, and peregrines are certainly willing and able to kill herring gulls.

The South West Peregrine Group explained that the 2014 gull chicks were found at a spot where peregrines used to nest. They wondered whether a local peregrine pair nesting nearby had lost their own eggs or chicks and had, being in full parenting mode, transferred their energies to the gull nest instead, driving off the actual gull parents in the process. How the two species got around the problem of their different chick-feeding styles is another one for speculation. The video showed the gull chicks taking bits of meat from the peregrine's bill, just like peregrine chicks would do, but herring gulls feed their young differently, by regurgitation, stimulated by the chick pecking at the red spot on the adult gull's bill. Perhaps the proffered chunk of red meat looked enough like a red bill-spot to trigger pecking behaviour from the gull chicks.

It's a pity that we could not know these baby gulls' fate after they fledged. They were certainly convinced that the peregrines were their mum and dad, but would they have grown up believing that *they* were peregrines too, and behaved accordingly? The peregrine kills its prey by stooping headlong on it from high in the air, reaching eye-watering speeds in its vertical dive before punching the prey out of the sky with a punch from the balled-up talons. Versatile though the herring gull is, I think that would be beyond its capabilities. And what about courtship? A herring gull trying to flirt with a peregrine would not get very far and quite possibly would get itself killed. It all depends on which natural behaviours are hard-wired and which are learned (and no doubt many are a bit of both).

The presence of gulls can influence aspects of other birds' behaviours. If you visit Inner Farne island in the height of the breeding season, you will be relentlessly assaulted by Arctic terns, and in between dodging their sharp bills you may notice that there are more tern nests close to the paths than further away from them. If you use the loo while you're on the island, you'll have to run the gauntlet of a small enclosed courtyard where there are particularly high numbers of tern nests. The terns react to you as though you are the most appalling dangerous monster, which is rather hard to take when all you want to do is admire their elegant beauty from a safe distance, but they still choose to nest in all the places where the island's human visitors have to walk. Why? The answer is that there are also gulls of various species nesting on the island, and these gulls are predators of tern nests. The closer the nest to a footpath, the less likely a gull is to go near it, because the gulls are rather afraid of people. The terns that nest near the paths might get very stressed by passing humans, but they are less likely to actually lose any eggs or chicks than the terns nesting further away.

As I mentioned before, the gulls on the Farne Islands don't just rob nests, but also rob other seabirds of the fish they have brought in for their chicks. Puffins are targeted relentlessly, and watching one lose its whole catch to a gull does make you wonder whether this is taking its toll on the puffin chicks' survival. However, studies have shown that its impact isn't significant, and moreover, that puffins that nest in the spots where there are the most gulls manage to bring in *more* fish, because the gulls are often preoccupied with squabbling among themselves.

That's not to say that gulls couldn't ever be a serious problem for other nesting birds. Here we wade into controversial waters. In England, all wild birds have legal protection and can't be harmed unless you have applied for and been granted a licence from the government. You won't be granted any such licence unless you have a very good reason (for example, you might need to remove a bird's nest from a bit of safety-critical equipment on an airfield). But there are also the so-called 'General Licences' which, at the time of writing, permit the killing of a very small number of wild bird species for certain specific purposes, without needing to apply for a licence first. Most of said purposes are to do with birds coming into conflict with human interests, and we'll be talking about those later in the book. But there is also a general licence which permits the killing (or nest destruction) of 'certain species of wild birds to conserve flora and fauna'. This licence covers only seven native British species, one of which is the lesser black-backed gull.

Lesser black-backs are nest predators, of course. Unlike herring and great black-backed gulls, which prefer high coastal nest sites, they often nest inland and on marshland, and they will grab eggs from the nests of birds like waders and terns if they get the chance. This isn't a serious problem in most circumstances. The waders and terns protect their nests either through camouflage (of nest, eggs and incubating adult bird alike) or by living in such big colonies that they can fend off most predators through sheer force of numbers. Some nests will still be lost, but not enough to do any harm to the local population as a whole.

The problem comes when numbers are so low that every lost nest is significant, and colonies are too small to mount an effective defence. This is when conservation managers might make use of the general licence and kill off some lesser black-backs and/or destroy their nests, in a bid to protect the nests of rare waders and terns and improve their breeding success. The RSPB has culled lesser black-backs on a few of its reserves for this reason, and has come under fire for it, because lesser black-backs are themselves on the Amber List of Birds of Conservation Concern. Elsewhere in England, the RSPB has campaigned vigorously to protect nesting lesser black-backs from being culled on the grounds of air safety, so it's not surprising that accusations of hypocrisy have been flying about.

The RSPB's business is wildlife conservation, so it has a serious dilemma to resolve when one rare species is negatively impacting the survival of another one. The way forward is to first establish whether there really is a problem (because, after all, predation is natural and often sustainable). If there definitely is a problem, then conservationists need to find ways to mitigate it without taking lethal action. Sometimes the solution is a bit on the labour-intensive and resource-expensive side. For example, the little tern colony at North Denes, Great Yarmouth, several hundred pairs strong, suffered heavy losses from kestrel predation recently, with just one pair of kestrels causing near complete breeding failure in 2005. In some years, voles (the usual kestrel prey) are much scarcer than others, and fluffy little tern chicks pottering about on the ground make a very suitable substitute, being about the right size and quite a lot easier to find and catch. The problem was largely solved by offering the predators an even easier option – pre-killed white mice and day-old chicks, left out on prominent perches for the kestrels to help themselves. Supplementary feeding could also work to divert gulls from other birds' nests during the breeding season.

Starvation is not a nice way to die, but is the fate of most wild animals that do not meet their end courtesy of a predator. Kittiwakes, wintering far out at sea and struggling to find the small fish they need, are starving in large numbers. For our more enterprising coastline gulls, willing to eat anything and everything, there's less chance of straightforward starvation, but more chance of meeting an unpleasant end while searching for food, and that risk is much bigger for gulls that forage in areas where there is human activity.

Those that forage on rubbish tips expose themselves to an unpleasant disease known as botulism, caused by toxins released by the bacterium *Botulinim clostridium* type C. The toxins made by *B. clostridium* are used to make Botox, a muscle relaxant that, when injected into a human forehead, makes that forehead become less wrinkly (though also less expressive). When consumed by gulls, the toxins cause muscle paralysis, particularly in the neck (giving the condition its alternative name 'limber-neck'). Botulism-stricken gulls look poorly and hang their heads and wings – eventually they cannot walk or stand. They often die from drowning, dehydration or starvation. They can recover – wildlife rehabbers can usually treat the birds by making sure they are given clean food (sometimes by tube if they cannot feed themselves) and kept warm while the toxins pass through their systems, and they are usually all better after a couple of weeks. However, botulism-stricken gulls in the wild are unlikely to recover, and botulism poisoning is a significant cause of population declines in several British gull species.

Foraging gulls can also get themselves tangled up or injured, by fishing gear and various bits of rubbish. The plastic 'collars' that hold sets of beer cans together are particularly notorious for entanglement possibilities (it's always a good idea to snip them before disposing of them). Unless they can free themselves, they will probably be handicapped to some extent by the encumbrance and much more likely to starve or be targeted by a predator. Gulls may also unintentionally include bits of rubbish among their nest material, which is a tangling risk for their chicks as well as themselves. Other new hazards brought into the gull's world by people include overhead wires, which are deadly if you fly into them full tilt. The same goes for wind turbines – it goes without saying that we need to be collecting our energy sustainably as much as possible, but turbines sited close to bird colonies or on important migratory routes can cause significant mortality.

As our wild countryside shrinks in extent and its richness is watered down, gull numbers are likely to continue to fall in rural areas. At present, their numbers are climbing in towns and cities. It's not yet enough to offset the wider losses, but maybe someday it could be. The plotline of our gulls' natural history story is beginning to change. Our attitude towards gulls is changing too, and it's hard to predict where this twist in the tale will take us.

CHAPTER 3

The Herring Gull

Herring gull

APARK IN Falmouth, Cornwall, on a bright winter's afternoon, warm enough (in coats) to sit outside the cafe by the lake. We're pretty near the coast, and herring gulls are everywhere. They are almost all young birds, mostly in their first and second winter. The light is luscious and I keep setting aside my mug of tea to reach for my camera. I track a young gull flying down to alight on a railing – for a moment it's angelic as its backlit wings sweep forward. On the water, the gulls are vaguely cruising around in search of food but are constantly diverted by non-edible playthings. I watch one larking about with a little plastic football, trying to grab it, trying to submerge it, knocking it out of reach and flap-paddling over to retrieve it. Another is playing catch with a fallen leaf. New arrivals skim low over the lake, touching the water's surface with a casual wingtip as they bank and turn. It's supposed to be a wide-ranging birding day and we have rarities to chase, but I could happily stay here and watch herring gulls all day.

It's a few years earlier. On a trip to the northern side of Cornwall, I visit the lovely town of St Ives. Here, after a morning of yomping up and down its steep side streets, I head down to beach level and purchase a warm and possibly award-winning Cornish pasty from a bakery. I then make the schoolgirl error of taking it down to the seafront to eat it, so I can admire the many gulls milling about in the area. I take a bite, and as I draw back my arm, a herring gull is suddenly there in the air right next to my face, pumping its wings and lunging its head pastywards. I have no time to react as it gets a good grip on my pasty with its bill, tears off about half of it, and makes its escape, while about fifty more herring gulls are already closing in. It's not often I'm intimidated by birds, but their utter fearlessness is unnerving, as is their size at this range. I let the gulls have the pasty.

Rewind quite a few more years, and I'm making my way home from the school train, along the seafront in Hastings. It's early autumn. On top of a parked car opposite the boating lake there is a juvenile herring gull, one of this year's babies so probably no more than five months old. As I get nearer, it patters across the car roof towards me, and when I pause to take a look at it, it jumps forward with a beat or two of its wings, and lands on my head. I'm not used to this sort of carry-on from our local herring gulls and for a moment I feel quite flattered and honoured, but then the gull begins sorting through my hair and raking my scalp, none too gently, and the wonder evaporates. With some difficulty I dislodge the young bird and walk on, briskly.

A little further back in time and I'm in the kitchen in our Hastings house, in winter, ignoring my family and staring out of the window across the patio and into the garden, as is my wont as a ditsy and bird-obsessed 12-year-old. One of our rooftop herring gulls flies down to land at the top end of the patio – an unusual event in itself. Then it struts over to my mother's treasured terracotta cat ornament, and begins to beat seven bells out of it. I make a mental note of this intriguing behaviour. My mother, however, is aghast. She throws open the patio doors and runs at the marauder, yelling at it to desist. Before she's halfway there, she slips over on gull poo and falls hard on her back. The gull watches, in disdain, before continuing its assault on the terracotta cat.

Later that year, our rooftop gulls have chicks, and they take against our youngest (actual, living) cat, Sam. We have three other cats, but for some reason they only have a problem with Sam. When he goes outside, they drop from the rooftop into the air and wheel overhead, before plunging down with furious battle cries. Sam flees in terror, back to the house. He was a confident and carefree cat before all of this, but before long, he's reluctant to go out at all. The gulls' campaign of intimidation ends only when their chicks have fledged.

So yes, I've known gulls to behave badly. You won't find many more enthusiastic advocates for herring gulls than me, but even I can't pretend they are consistently charming and delightful. Ask a few other seaside citizens how they feel about the gulls and you'll quickly hear some very strong views. Peruse the letter pages in local newspapers and it won't be long before you find one angrily calling for a wholesale cull of the 'seagull menace'. The birds' crimes include (but are not limited to): crapping copiously over cars and garden furniture; tearing up bin bags and strewing the contents across the streets; attacking people (and pets) that go near their nests; stealing unattended food; keeping everyone awake with their all-night screaming parties; and being very big and alarming, which frightens the children. As a (non-frightened) child, I wrote a letter to the Hastings Observer defending the gulls, and explaining how pointless any local cull would be anyway (other gulls would simply take their places). It was printed, with my age (9) in brackets. No one listened, of course.

This controversial gull, causing havoc in seaside towns throughout the British Isles, is *Larus argentatus*, as you might remember from chapter 1, and if you want to give it its full name, it's *Larus argentatus argenteus*. According to the *Handbook of Birds of the World*, there are only two subspecies of herring gull. *L. a. argenteus* is our one, and as well as Britain it can be found in Iceland, on the Faroe Islands, and from western France to western Germany. Some of these birds, especially the ones living in more northerly areas, move south for winter, passing the colder months in sunny Spain and Portugal. The other subspecies is the slightly bigger and slightly duskier-coloured *Larus argentatus argentatus*, which breeds from Denmark north-eastwards across Scandinavia. Some of these birds also winter further south and west, including in Britain, where they hide among groups of *argenteus* herring gulls, the young ones in particular offering a stern challenge for the larophiles.

My sole experience with the north-eastern subspecies, *L. a. argentatus*, comes from a single and slightly terrifying boat trip out of Bridlington in north Yorkshire, on a grey and blustery October day. The trip was supposed to provide us with sightings of skuas and shearwaters – relatively glam seabirds which most of my fellow passengers were very keen to see. The waves rocked us about with increasing vigour as we made our hesitant way further out into the North Sea. From the back of the boat, buckets of chum (fish guts and other malodorous gunk) were being dispensed across the choppy water to attract the birds. Soon a sizeable maelstrom of gulls had gathered, but nothing but gulls. The skipper tried to rally everyone's enthusiasm by pointing out a herring gull of the subspecies *argentatus*, among the swirl of ordinary *argenteus* herring gulls and also great black-backs that were following us along. I clung to the siderails and knelt up on the seat to get a better look at the northern wanderer. It was, indeed, quite distinctive with its hefty build, ferocious expression and rather dark grey wings, with less black coloration on the outer parts of its flight feathers than the many *argenteus* birds had. A few others just like it were soon tagging along, and I set about cataloguing their distinctive traits in my mind. Meanwhile, a small posse of gannets turned up and did some exhibition plunge-diving into the sea around us at point-blank range, doing their best to make up for the continued absence of skuas and shearwaters. A storm warning chased us back to land ahead of our allocated time, and the *argentatus* herring gulls left us to it, remaining at sea while their local *argenteus* cousins followed us back into the safety of the harbour.

The adult herring gull of our own resident subspecies is pretty easy to identify. It's big – not as big as a great black-back but still a reasonably formidable creature. It is a bit larger than the lesser black-backed, and bulkier too with proportionately shorter wings, while the lesser black-backed looks slimmer and more attenuated at the rear end. Its back and wings are a light silver-grey, paler than any of our other bigger gulls – the yellow-legged is a couple of shades darker, the lesser black-backed much darker, and the great black-backed almost jet-black. The outer flight feathers are black at their far ends, with small white tips, and additional white spots or 'mirrors' near the tips as well on the very outermost couple of feathers. This is only really obvious when it flies, and that's also when you'll see a narrow white leading edge and slightly thicker white trailing edge to the whole wing. The plumage is otherwise pure white, though in early winter, in its freshly grown post-moult plumage, it has a little bit of greyish streaking on its head and neck. By early spring, these grey bits have worn away, leaving a pure white head, and wear and tear has often also reduced or removed the white tips to the flight feathers. Sun bleaching gradually fades the black parts of the feathers to a dull blackish-brown too, so by June, with the annual moult about to begin, most gulls are looking more drab and also quite raggedy. From late spring they begin to show gaps in their wings as flight feathers are shed and regrown, the inner primaries and outer secondaries falling first, and the gaps moving along the trailing edge of the wing as moulting progresses. Tail moult begins from the tail centre and heads outwards, so herring gulls in early moult can look a little fork-tailed.

The adult bird's bill is yellow, with a red spot on the chunky pointed bit on the underside of the lower mandible. Its eyes are pale yellowish, giving that rather cold and unfriendly facial look, and the ring of skin around the eye is bright yellow. The legs and feet of the adult herring gull are a pleasing light pink, a bit brighter in the breeding season. Their representations in art and ornament that you'll find in seaside gift shops almost always have *yellow*, not pink legs, which annoys me – how hard is it to look at an actual herring gull and get it right? Maybe they paint the bill first and then can't be bothered to wash the yellow off their brushes. Actually, some eastern *argentatus* individuals do have yellow legs, but this hardly ever happens in the subspecies *argenteus* – the big gulls with yellow legs that you see in Britain are not herring gulls, they're either lesser black-backs or (much less likely) yellow-legged gulls – both rather dark grey, rather than silvery like the herring gull.

Younger herring gulls are harder to identify, especially those in their first year, before the silver-grey back starts to appear. The youngsters of summer, the ones that used to fall off our roof on a regular basis, are entirely darkish muddy grey in their brand new juvenile feathers, with darker marbling and streaking. Their big eyes are very dark brown (making them look a lot cuter than their glassy-eyed parents), their bills are black, and their legs are dull pinky-grey. They look very like young lesser black-backs, but are a bit paler and more variegated, and their shape is also a bit different. They're more stocky and barrel-shaped with shorter wings, while lesser black-backs are slightly smaller, slimmer and with long wings, making them look rather more graceful and elegant. These differences in shape are, of course, still there in older birds, but by then there are more obvious plumage differences as well.

Juvenile body plumage is soft and not very durable, and in autumn, the juvenile gulls moult these body feathers (though not the larger wing feathers – those are kept until the first full moult a year later). They grow new body feathers, and now they are no longer juveniles but first-winters, as this new body plumage will see them through their first winter.

The fresh, first-winter plumage is pretty similar to juvenile plumage. The main difference is that the background colour is a bit paler, with the darker markings standing out more. At the end of winter, some of the head and body feathers get replaced with another set, a shade paler again, in a partial moult, and the birds are now in first-summer plumage – though you need to look closely to see the difference. Then, in late summer, when they are a year-and-a-bit old, they go through a full moult for the first time, finally replacing all their worn-out, juvenile-aged flight feathers for the first time, as well as replacing their body feathers for the second or third time. They are now in second-winter plumage. And, once again, at the end of winter they have a partial moult of head and body feathers, and they are now in second-summer plumage.

And so it goes on – a full moult each late summer, a partial moult each late winter, and each time they grow a more adult-like set of feathers. Their bills, legs and eyes also gradually change to more adult-like colours. Their eyes get paler, their legs get pinker, and their bills get yellower. They eventually look completely adult at the age of four or five. For adult herring gulls, summer and winter plumage are not very different, but winter-plumaged birds have a slightly streaky head. This is replaced with pure white feathers when the bird goes through its partial moult in late winter.

Often, herring gull flocks in winter will hold birds of all ages – adults, first-winters, second-winters, third-winters, and fourth-winters, which are often indistinguishable from adults. You don't need to be *too* gull-nerdy to work out the age of most birds you'll see at this time of year. The first-winters will be dark-eyed, dark-billed and speckled all over. The second-winters will still be quite speckly but will have a mostly plain silver-grey back (or saddle), and often will have pale eyes and a partly pale bill. Third-winters have a mostly white head and body, usually pale eyes and a yellow bill, but quite a bit of obvious brown speckling in the wings and on the tail-tip. Fourth-winters look like adults in almost every way but usually have a little tell-tale trace of brown in the wings and/or tail-tip. However, every individual is different, and matures at a different rate... and moreover each *bit* of it matures at a different rate – one might develop an adult-like bill quite quickly but retain juvenile-like eyes for longer, while for another it could be the other way round. A group of young gulls of exactly the same age could all look quite different from one another, and one month on they'll all look a bit different to how they did before. Therein lies the joy and frustration of gull identification.

I need, at this point, to talk about the words we use for gulls' (and other birds') ages, and plumage types. The system you'll hear most often is the one I've been using above – there are juveniles (birds in their first set of feathers, which haven't had their first moult yet), and there are adults (birds which have reached their final plumage type

Herring gull

and will not change in appearance from year to year any more). In between, there are first-winters, first-summers, second-winters and so on. But these terms aren't that useful. At almost any given time, a subadult gull will be moulting at least some of its feathers, and what we call a first-summer bird is actually wearing a mixture of juvenile feathers, first-winter feathers and first-summer feathers. Add to that the aforementioned vast amount of individual variation in appearance and moult timing, plus disagreement about when summer and winter actually start anyway, and you can see why many gull enthusiasts now choose to opt for a different and simpler method of describing age. The 'calendar year' or 'cy' system just ages the bird based on the year it was born. A 1cy bird was born in the current year. In December, all young gulls born in that same year are 1cy birds, but on 1 January they all instantly become 2cy birds. Here and now, in March 2019, there are no 1cy herring gulls around because it's too early in the year – none have been born yet. Those born in 2018 are 2cy birds, those born in 2017 are 3cy birds, and those born in 2016 are 4cy birds. Because nearly all northern hemisphere gulls are born in April or May, this system allows you to give the age of any gull pretty accurately (as long as you have a handy calendar to check which month you're in) while neatly bypassing any need to attempt to describe its actual plumage.

The biology of the herring gull is pretty typical of the 'big gulls'. The adult pairs breed once a year, with pairing-up and nesting activity beginning in early spring if not sooner. Through April they are building nests and beginning to lay eggs. These hatch in late spring. The chicks are born bright-eyed, fluffy, and distinctly adorable. They are able to stand up and wobble along a bit from their first day of life, but need their parents to bring them food. They take several weeks to grow big and feathery enough to fly well, and their parents continue to feed them regularly for several more weeks, sometimes much longer. In the later weeks of the breeding season and for several more weeks beyond, the main annual moult takes place, during which the juvenile birds moult their body feathers and all the older birds moult the lot, flight feathers included. Because the flight feathers are shed and regrown gradually, the birds never lose their ability to fly, but this makes the moult a slow process. It is more or less finished by late autumn, leaving all the gulls resplendent in a complete fresh set of feathers. They are now ready for the winter, which they will spend doing little but eating, hanging out, and perhaps engaging in a bit of light territorial defence and flirting.

Another activity that herring gulls engage in is play. It's long been observed that they will drop hard-shelled prey objects, like shellfish, onto rocks or other hard surfaces to break them, but young gulls in particular will also spend time dropping an object (which may or may not be edible) in mid-air, and then swooping down to catch it before it hits the ground. And, as I described at the start of this chapter, they'll do the same thing on the water, dropping and retrieving a floating object. Play seems, on the face of it, pretty pointless and a big waste of energy, and if it's freezing cold and there's not much food around, playful behaviour is rare. But when there's a bit of wiggle room in a gull's energy budget, time spent in play behaviours isn't a waste at all. It's a safe way of practising ways to hunt and catch prey, honing one's speed and skill at making a grab and a catch, and it may also lead to innovative new behaviours that open up new ways to exploit food sources.

The herring gull doesn't usually become properly mature and ready to breed until it's four years old. That's why, in summer, most of the gull flocks you see away from breeding areas are composed of younger birds – the adults are busy with their nests. However, you may see two- and three-year-old birds at the breeding colonies too, checking out possible nest sites and perhaps getting romantic with other gulls. It pays these young ones to have some sort of clue about what's involved with pairing up, finding a nest site and raising a family before they actually dive in and give it a try. Seabirds in general live for a long time and mature slowly, so it's pretty common to see young ones loitering around the breeding colonies, in search of tips and guidance. In urban colonies particularly, birds in their third and even second year will sometimes actually breed.

Young male herring gulls tend to head back to their natal colonies when they become interested in breeding, while young females are more likely to wander around and eventually join a different colony which could be many miles away. When they start to spend time at colonies, the young gulls will be keeping tabs on the established pairs, ready to make overtures towards any older, successfully breeding gull that becomes suddenly single. Young birds will often pair up with each other simply because there are more of them around than single older birds, and young males will try to establish a territory of their own rather than muscle in on someone else's, but it's a better bet, when you are a first-time breeder seeking a mate, to try to attract someone with experience. Such opportunities are rare, though. Death or unexplained disappearance aside, pair bonds between herring gulls are long lasting. They're as devoted as swans, but less renowned for it because they don't show their affection through slow dancing across picturesque lakes together and making heart shapes with their necks, but by vomiting up a pile of semi-digested fish (and perhaps also chips), when prompted, for their beloved to eat.

Gulls, like other birds that stay together for life, have lots of courtship behaviours, used to demonstrate their qualities as a partner and their commitment to the breeding season ahead. They perform these behaviours when first getting to know one another, but also go through the routines again each year as the breeding season approaches, to renew their vows, as it were. They also have ways, vocal and visual, to signal other things to other gulls, such as 'get away from the vicinity of my nest, NOW' and 'I'm so sorry, I didn't realise that was your nest, please excuse me'. Approaching another gull in a head-down, horizontal posture, with the wings slightly held out, indicates aggression and the other is likely to back off. If there is a confrontation, the birds most often fight by grabbing at each other's wings and wrestling. Really violent fighting is rare in gulls – they are naturally social after all, so are good at defusing social tensions. You're much more likely to observe full-on bloody battles between apparently innocuous (but territorial) little birds like blue tits and robins.

Social interactions begin in the nest, between adult and youngster, when the newly hatched and newly hungry chick begs for food by pecking at its parent's bill while whistling feebly. This stimulates the adult to throw up a bit of food, which the chick then eats. Later on, when the chicks are bigger, stronger and able to run about, they do the same thing but with more vigour and a louder, squealier whistle. The animal behaviour biologist Niko Tinbergen carried out a famous study on the begging behaviour of baby herring gulls, observing that they direct their food-soliciting taps at the red spot on their parent's bill. He then discovered that you don't even need to have an adult gull present to trigger the begging behaviour – offer the gull chick a stick, painted yellow with a red spot, and it will do the same thing. A black spot also works, suggesting that it may be the tonal contrast between background colour and spot that gets the chicks interested, rather than the actual colours.

As an aside, another notable discovery about apparently hard-wired gull behaviour, but in adults in this case, arose from Konrad Lorenz's work on 'supernormal stimuli'. He noticed that herring gulls incubating on the nest seemed to favour some of their eggs over others. If the eggs were rolled out of the nest, the parents would invariably roll the biggest, spottiest egg back first. Studies have since been carried out to see what would happen if incubating gulls were shown an egg much, much bigger and spottier than any that nature could devise, using 'dummy eggs' made of plaster and painted to be like extra strikingly patterned, intensely coloured herring gull eggs. The gulls reacted very strongly to the artificial monster eggs, pushing their own eggs aside as they tried to drape their bodies over the magnificently huge fakes. It's easy to see this as a failing in the herring gull's psychology. Not only are they clearly not very bright, but they seem to have a certain shallowness, to be so readily beguiled by such wildly exaggerated fakes. And yet, advertising agencies do exactly the same thing when selling products to us humans, often with great success, so who are we to judge?

Herring gulls make great parents, at least for their firstborn and their second-born. The complete clutch is nearly always of three eggs, and usually three chicks will hatch, but the chick from the third-laid egg is at a serious disadvantage right from the start. The third egg is usually clearly smaller than the first two, and it also begins 'life' later, as the first two eggs have already been incubated for a couple of days by the time the third one is laid. So by the time chick three hatches, it might be in trouble if chicks one and two are fit and healthy. They have already had a couple of days of growth and will easily shove their little sibling aside at feeding time, so it gets fed last. Any shortage of food, and chick three, the 'insurance chick', will be first to starve. In some gull species, the parents may even help hasten their last-born's demise, with deliberate neglect and actual attacks. There may also be battles between the chicks themselves. Parental infanticide and siblicide is less unlikely if food is plentiful, though, and among urban herring gulls it is more common for all three chicks to make it to fledging age than for chick three to succumb.

Where do our herring gulls breed, besides on seaside town rooftops? They are among a small group of bird species which naturally nest on sea cliffs and safe, flat islands and beaches, and have found that rooftops make a pretty good substitute for those things in all the ways that matter. Like peregrine falcons, house martins and rock doves, herring gulls like to nest on buildings as it puts them safely out of the reach of mammalian predators, and (often) still within easy reach of a food source. Natural nest sites need to offer the same deal, and so you'll mostly find herring gulls nesting on high, sheltered cliffs and rocky outcrops. Another way to escape the foxes, rats and other furry nest raiders is to nest on small islands, and here it doesn't matter if you're on a high, rocky promontory or a flat beach – they're equally safe. Some of the Hastings herring gulls nest on the sandstone cliffs that mark the eastern edge of the town, rubbing shoulders with fulmars and jackdaws.

In April 2017 I went to Havergate Island to look for brown hares, which are famously prolific there in the absence of foxes, dogs and other dangers, and also notably relaxed in human company. This is a small and very low-lying shingle and saltmarsh island in the River Ore in Suffolk. It's only a (strong) stone's throw from the mainland but the river channels surrounding it flow fast and deep enough to deter most mammalian explorers (the hares were brought there by people, centuries ago). Exploring the island with a dozen or so other day-tripping wildlife watchers, I headed for the shingle bank where I'd seen several hares on a previous visit. That had been lateish autumn. Now it was mid spring and the shingle bank looked very different, because back then it had been bare, and now it was entirely covered with gulls. The same was true of the large, shallow lagoons that covered a large area of the island's surface.

Most of these gulls were lesser black-backs, but I'd guesstimate that there were at least 30% herring gulls here, and counting both species together there must have been a few thousand pairs present. Some of those on the shingle were nest building, others just standing about. The pairs with nests were quite well spaced but it was clear that this was a colony, nesting (quite) close together and interacting socially as and when they needed to. On one of the lagoons, I noticed a herring gull and a lesser black-backed perched closer together and more comfy looking with the proximity than you'd expect for mere acquaintances … mixed pairs and hybridisation between these two species have been observed here. The island has a beach shoreline, onto which the tidal river must deliver plentiful scavengeable food for the gulls, and it's just a short flight further to the actual sea coast and the many fishing boats offshore. Life is pretty easy, for now. But Havergate Island is extremely vulnerable to sea surges, given its position and lack of height. The storms of winter 2013 caused major flooding, nearly wiping out the brown hares and doing lots of damage to the habitat as a whole, and events like this are likely to be more frequent in the future. So this gull colony may not have much time left.

There are herring gulls nesting on the Farne Islands, off Northumberland, as well. These islands are of a very different character to Havergate – they're lumpy, mostly rather sparsely-vegetated granite chunks in the open sea, with grey seals loafing on their more accessible shores, and (in spring and summer) lots and lots of seabirds on the cliff sides and tops. In June 2014, I boarded a little boat at Seahouses Harbour, and we crossed a short stretch of water towards one of the Farnes – Staple Island. As we approached, we chugged past larger and larger groups of floating puffins and guillemots, and watched shags diving for their prey like miniature porpoises, while kittiwakes lifted off from the low, shoreline rocks, carrying clumps and clots of seaweed for their nests.

On the island itself, auks, shags and kittiwakes were nesting in profusion on the rocky parts, while up on the grassy summit there were a few herring gulls and a few lesser black-backs. They stood sentry in pairs near their nest sites, casting their cold gaze across much of the island. The puffins also nested on the grass-covered tops, their chicks safely tucked away in burrows, but the adult puffins were not safe from the gulls' attentions. When a puffin came whirring in from the sea with a load of sand eels clamped in its colourful bill, it would pitch down and then sprint to its burrow with a gull or two in hot pursuit, snapping at the fish. Often the gulls managed to appropriate some of the puffin's catch before it could complete its dash to safety. Later on, I noticed a herring gull making its way up from one of the sea cliff sides, with the pathetic little corpse of a newly hatched guillemot chick dangling from its bill. There's no shortage of gull food on this island – but its resident auks certainly would have good reason to write to their local newspaper with herring gull-related complaints.

Gulls nesting on sea coasts elsewhere don't necessarily have such easy access to food as the Havergate gulls or the Farnes gulls. Only where there is a good food supply within easy reach of good nesting habitat can a dense gull colony form – but there might be room for one or a few pairs, or a smaller and more scattered colony. Lone nests are more vulnerable, though. Even within colonies, those nests on the outskirts with few near neighbours are the ones that suffer highest predation – including from other herring gulls. There is even evidence that a few colonial-nesting herring gulls actually specialise in cannibalism for a short part of the breeding season, and they make a point of attacking the most isolated nests in the colony, grabbing chicks that are just a few days old. So the best way to improve your chances of breeding success is to be near the centre, with lots of neighbours – the chances are that the majority of the neighbours will not eat your babies, and that they will help discourage others from doing so.

Once herring gull chicks are more than a couple of weeks old, they are a lot less likely to fall victim to any predator, whether a marauding fox or that nice couple from the east side of the colony. They can now run quite fast, give a reasonably discouraging scream, and peck with vigour. Those that hatched from nests on flat ground will wander quite freely, though may be attacked if they go too close to other birds' nests. For cliff-nesters, there's less scope to explore, and those who are too adventurous may fall to their doom. If they just fall a short way, though, and land unhurt on a different ledge nearby, their parents will probably simply continue to feed them as if nothing had happened. A family of three chicks could end up living in completely separate spots in the weeks before they fledge.

Herring gull eggs take about 30 days to hatch, and then the young another 35–40 days to make their first flights. Add on the courtship and nest-building period before all of this, and at least 30 more days of feeding the young ones after they have fledged, and the breeding season takes up an intense four months of the year for adult herring gulls. This period begins and ends at close to the same time for all the pairs in the colony, at least in larger colonies. Having a well-synchronised breeding season helps improve breeding success, as it means all the little vulnerable chicks are around at the same time, in such numbers that even the most committed predator couldn't eat them all. The months that follow breeding are concerned with getting through the moult, and then surviving winter – but (for the adult gulls) merely surviving isn't enough – they need to do well enough to be in good condition and ready to breed again the following year.

When not nesting, herring gulls follow the food. For some adults, that means no change is needed – they can continue to spend their time mostly around their nest site, and head down to the strandline to forage when they need to. For others, winter is time to roam, and we are learning more and more every year about the wanderings of British – and European – herring gulls. Much of this new information comes thanks to ringing projects which use Darvic rings. These are large rings bearing alphanumeric codes that can easily be seen from a distance. This means that the researchers can track birds' activity without the need to catch them to read the ring number. Lots of larophiles have a particular interest in logging the Darvic-ringed birds they see, especially when it's a bit of a slow day at the rubbish dump. This is a real boon for the researchers.

In July 2013, I was in Hastings to visit my father and took a stroll with my camera along the seafront. At some point, as an almost-adult herring gull drifted overhead, I took a photo which didn't come out particularly well. Later, checking the images, I was about to delete it when I noticed that the bird was wearing a bright orange Darvic ring. My photo showed the ring number clearly – U4BT. I found out online which project the ring was from – the North Thames Gull Group – and sent them an email with my photo of the gull. They quickly replied, sending me a link to their website, and a page all about the gull in question. U4BT was, I learned, caught and ringed in February 2012, one year and five months before, at the Pitsea landfill site in Essex. It was recorded then as being in its third calendar year. My sighting was the first one of the bird since it had been ringed.

I've just re-found that email reply, and I'm now clicking the link again, five years on, to see whether U4BT was ever seen again. I am happy to report that it has indeed been observed by others, not once but three more times, the most recent just five months ago. In March 2016 it was seen again in Hastings – the timing suggests that it was thinking of breeding there. Two more sightings followed – in February 2018 it was at the Bexley rubbish tip in Crayford, Kent, and it was there again in June 2018. The last sighting is a bit of a puzzle. Why was it there in June, mid breeding season, when the 2016 observation suggested it was interested in breeding in Hastings? We know that herring gulls are loyal to their nest sites. So perhaps it had prospected for a nest site in Hastings in 2016 but been unsuccessful, and relocated further north. Or perhaps it is indeed nesting in Hastings, but considers that Crayford is not too much of a commute from there. Or perhaps for whatever reason it has not yet managed to do any breeding at all, even at eight years old. Future Darvic ringing studies will help us to solve riddles like this.

I decide to wander around the North Thames Gull Group's website a little more, and check out some other herring gulls' movements. I read about R4BT, a first-year gull ringed in 2012 in Essex, which went off to Calais in northern France a few days later and is still there now in 2018, a couple of short trips to south-eastern Kent notwithstanding. Another, LU4T, was ringed as an adult in 2011, and since then has apparently been unable to decide whether it preferred to winter in northern France or Leicestershire, while M3TY is a Norwegian breeding bird, spending its summers in the most north-easterly part of Finnmark and then travelling more than 2,500km south-west to tour the rubbish tips of south-eastern England through the winter months.

Observations of ringed birds (and recoveries of dead ringed birds) show us that the herring gulls that live in or visit Britain are not at all averse to a bit of travel. Satellite-tracking studies cast a more nuanced light on this, and highlight how much variation there can be, even among birds all nesting at the same colony. Four GPS-tracked adult herring gulls that were nesting in the Cornwall town of St Ives showed remarkable disparity in how they spent their off-duty time. Peter Rock, a very experienced gull researcher, and his colleagues tagged the birds (two males and two females) in May 2014 and followed their activities through the summer. Two of the four roosted at their nests, but another sometimes slept out at sea and the fourth was awake and wandering offshore through most of the night. Two of the birds never wandered far from their colony but the other two regularly made flights of 100km or more. Between them, the four covered some 32,000km through the summer, and that is when they had chicks to tend. In winter, without this limitation, the potential for travel is even greater.

The BTO's ringing recovery data shows that 969 herring gulls ringed in Britain have reached France, 678 have reached the Netherlands and 133 have turned up in Belgium. There are some much longer journeys recorded too – birds ringed here have been found in Morocco, Russia, Iceland, and almost every mainland northern European country with a coastline. Then there is the gull ringed as a chick on the Bass Rock, Scotland, in 1965, which was found in Greenland six years and nine months later – it had crossed 2,576km of mostly open sea. Another great traveller moved 2,633km between Essex to Norway. This bird, not content with making one of the longest herring gull journeys on record, is also aiming for a longevity record – the Norway sighting took place 30 years, two months and 27 days after it had been ringed as a first-year bird in Essex. The current longevity record for a British-ringed herring gull is 32 years, nine months and 25 days, but our Norwegian explorer could still be around today and in with a chance of beating that impressive lifespan. It'll have to stay alive a little longer to set a new European longevity record though – a Dutch bird made it to 34 years and nine months old.

Even including migratory wanderings, though, the total world range of the herring gull isn't very big – this bird is pretty much confined to northern Europe. The population size in total at the time of writing (2019) is estimated at somewhere between 1.37 and 1.62 million birds. That's less than 0.00025 herring gulls for each human on Earth. In the UK, our breeding herring gull population is

140,000 pairs. So 280,000 adult individuals, sharing their nation with 66 million humans. Just over 0.004 herring gulls per person. Or, to look at it the other way round, each UK-based herring gull has 235 humans to annoy.

When you live in a seaside town it's easy to get the impression that breeding herring gulls are a) numerous and b) thriving. Let's look at point a) first. The most numerous breeding bird in the UK is the wren – we have about 17 million of those in 2019. The house sparrow, which has famously suffered a massive population crash over the last few decades, still tops 10 million. Plenty of other less familiar birds have UK breeding populations of much more than 140,000 pairs – among them the meadow pipit (2 million pairs), the chiffchaff (1.2 million pairs), the guillemot (950,000 pairs) and the fulmar (500,000 pairs). The herring gull isn't even our most abundant breeding gull species. It's outnumbered by the kittiwake (380,000 breeding pairs), and its population is more or less equalled by the black-headed gull.

What about the thriving part? Well, there's a two-part answer. Herring gulls in urban areas have been increasing. We're not sure by how much, but numbers nearly doubled between 1993 and 2002. But these gulls still make up only a small proportion of the total UK population (about 14% at the last count). Most of our herring gull population breeds away from towns, on remote coasts and islands. And these gulls are not doing well at all. There has been a long-term and serious decline, which is why it is on the Red List of Birds of Conservation Concern, and why it is the subject of a government Biodiversity Action Plan (BAP), aimed at improving its fortunes.

Looking more closely at the species' population demographics and trends in the UK reveals plenty that's of interest. We have this data thanks to various surveys, including three national censuses of breeding seabirds around the British coastline (Operation Seafarer in 1969–1970, the Seabird Colony Register in 1985–1988, and Seabird 2000 in 1998–2002). As I write, researchers are gearing up for the 2019 and final season of a fourth national seabird survey – Seabird Count – which began in 2015. There is also the annual Seabird Monitoring Programme, in which researchers supply population counts and figures for breeding success (or failure) from 25 key seabird colonies – this one has been on the go since 1986. Herring gull numbers are also collected by some of the volunteers who carry out the BTO's Wetland Birds Survey (WeBS), which monitors numbers of birds through the non-breeding season at wetland sites and has done so (in one incarnation or another) since 1947. Another long-running BTO survey, the Nest Record Scheme, also gathers data for herring gulls – in this case the likes of clutch size, hatching success, survival of chicks, and fledging age. The BTO has also organised winter gull roost surveys in some years, since 1953. The RSPB's annual and vastly popular Big Garden Birdwatch survey also collects data each winter on herring gulls (along with all other birds) that visit gardens, providing a broad view of how herring gulls are doing in urban environments.

Mash together all of this data and we can build a picture of where our herring gulls are living, how successfully they are breeding in different areas, whether their numbers are heading up or down in different parts of the country and in different habitats, and so on. The overall population trend results are that our total herring gull population declined by about 48% in 1969–1970 and 1985–1988, with a further decline of 13% from 1985 to 1988, and between 1998 and 2002.

The Seabird Monitoring Programme's data for the herring gull's breeding populations at 25 non-urban, coastal UK sites show a wiggly, but overall downward-trending line from 1986 to 2016. In more detail: populations fell steeply from 1986 to 1992. Then there was an increase between 1992 and 2000, eventually returning back up to 1986 levels, but then a rather steep decline for the next 12 years, to a new low point by 2012. A short-lived upward spike occurred between 2012 and 2014, but by 2016 numbers had dropped back to 2012 levels.

The other part of the SMP's remit – data on breeding productivity – shows a much more erratic line but here, too, the overall trend is downwards and the most recent data, from 2016, reveals an all-time low. On average each pair in this year fledged just 0.3 chicks, while in 1992, the previous population low ebb, the productivity was nearly one chick per pair. As we know, from 1992 the herring gull population began to climb again, and it seems reasonable to suggest that the increase was due (at least in part) to this improved productivity. However, things are different with the 2016 data. With productivity now so low, the downward population trend is not likely to be reversed in the near future.

Wintering herring gulls, many of which are not our own breeding birds but visitors from the continent, have apparently also suffered considerable declines. The Winter Gull Roost counts show that the crash has been most dramatic in northern and eastern Scotland and Wales, where numbers have dropped since the 1980s and 1990s and continue to drop. The counts also show increases in parts of the Midlands and southern England since 1993, but that these followed a steep long-term fall from 1963 to 1983.

Data like these help our government and conservation bodies to make decisions about how to manage our wild bird populations. Catastrophic declines and even local extinctions of some wetland birds in the second half of the 20th century drove the RSPB and The Wildlife Trusts to buy up and restore swathes of east coast marshland. Their efforts have allowed the return and spread of the lovely avocet as a British breeding bird, and bitterns, marsh harriers and bearded tits to bounce back from dismally low numbers. The severe decline of the house sparrow and the starling through the late 20th century prompted the UK government to remove both species from the General Licence in 2005. Before that date, any individual could kill house sparrows and starlings on their land without seeking special permission, provided they could show (if asked) that the birds posed a threat to important other interests, such as public health or livelihood. Both house sparrows and starlings were also placed on the Red List of Birds of Conservation Concern in the UK – a designation which automatically makes them a high priority for conservation efforts of all kinds.

The herring gull was added to the Red List in 2002, and the most recent assessment in 2015 indicated that it should remain there. What this has meant in terms of actual conservation is that the herring gull is, as of 2010, no longer listed on the General Licence, and enjoys the same full protection that nearly all other wild birds do in the UK. Any landowner having a problem with herring gulls must now apply for a special licence, which may or may not be granted. The lesser black-backed gull is still on the General Licence and landowners can kill it without applying for permission at all, for any of the following reasons: to prevent serious damage or disease, to preserve public health or public safety, and for conservation purposes (where the gulls threaten an endangered species). But none of these apply now to the herring gull, though it is still permissible under General Licence to destroy its nests and eggs to preserve public health or public safety.

Anyone looking at this with a cynical eye will conclude that no amount of protection of the adult birds is any use if people can destroy their nests and eggs at will. However, the herring gulls breeding in the wider countryside – the ones that are declining so worryingly – are not realistically going to be causing any public health and safety issues. This clause of the General Licence will be (and is being) utilised on urban-nesting herring gulls (but see footnote). They might still make up a minority of the UK gull population, but they are on the increase, and dramatically so.

I'm looking at a paper published in *British Birds* magazine in 2005, by Peter Rock. 'Urban gulls: problems and solutions' looks particularly at the herring and lesser black-backed gull population in towns around the Severn estuary, which increased from about 2,600 pairs to nearly 24,000 pairs between 1994 and 2004. The author calculates that, should numbers continue to rise at this rate, the region would hold more than 200,000 pairs (counting both species together) by 2014, which would extrapolate across the UK to more than a million pairs nesting in urban areas.

Natural England announced on 23 April 2019 that the General Licences under which gulls fall are revoked, with immediate effect. A new licensing system is likely to come into effect but at the time of writing, we don't know what it will be – check the Natural England website for the latest on this (www.gov.uk/topic/environmental-management/wildlife-habitat-conservation).

We know that this hasn't happened – the total populations of herring gulls and lesser black-backed gulls today in the UK are about 140,000 pairs and 110,000 pairs respectively, so less than a quarter of a million pairs combined, and that's across all habitats, not just urban areas. But herring gulls are still increasing in urban areas and are spreading to new towns all the time. These gains may end up offsetting the losses in the wider countryside – but we're not all going to be happy about that. Most people would rather have the gulls where they can't see them. And many of those people might also ask: what do we really stand to lose, anyway, if the herring gull were to carry on declining in the countryside *and* we were to curtail its population increases in towns? Would it really be to anyone's detriment to have a much smaller UK herring gull population than we did in the 1970s?

This is a point that's hard to argue. Really reliable, large-scale bird counts in Britain and Europe have only been going on since the mid-20th century, but everyone seems to agree that the relatively recent decline of our gull species is not the only significant population change they've undergone over the last few centuries, and many of the historical changes we've made to their environment have been very much to their advantage. The considerable expansion of the fishing industry, and creation of large, open rubbish tips, are two key factors that allowed their numbers to grow dramatically through the early 20th century. It's estimated that numbers in the Netherlands tripled between 1925 and 1940, from about 10,000 to 30,000, and in coastal Germany they doubled between 1930 and 1960. On Skokholm island off the coast of Pembrokeshire, a fairly typical 'wild' breeding ground for herring gulls, but within easy reach of fishing towns on the mainland, the population was logged at 250 pairs in 1928 and 1,350 pairs in 1969. So the declines in herring gull numbers that we've seen recently, brought about by our own actions, have affected populations that were already inflated, perhaps quite a lot, by other things we did before.

Working out what a 'natural' population level should be for any wild species today is a bit of an exercise in futility, as well as arrogance. We don't live in a natural world any more, if by 'natural' you mean 'not massively messed about with by humans'. When it comes to conservation in the modern world, we have to look at what we've got, work with what we have, and (arrogance again) figure out what we need and, to a certain extent, what we want. And conservation efforts are, and always shall be, in competition with other endeavours. With shrinking wildernesses and ever-expanding cityscapes, with ever more humans demanding more and more urban amenities, and having less and less tolerance for the mess, noise and chaos that nature brings into these parts of the world we consider to be 'ours', is there any chance that we can find a way to live at peace with the gulls?

CHAPTER 4

Gulls and People

Lesser black-backed gull

A N EARLY MORNING, down on Hastings seafront. Hardly anyone's around. The sun is just breaking over pink strands of cloud along the eastern horizon. When I visit my father, I often get up at first light, long before he's going to wake, and take a walk. Sometimes I climb the hundred (or at least it seems like a hundred) steps from Tackleway to the summit of the East Hill, alongside the near vertical path of the funicular railway from the seafront. Up here I can see across the whole of the Old Town, and the rooftop town of the gulls. Today, though, I'm heading for the seafront. I walk along the weatherbeaten boulders of the harbour arm as far as I can go and look down into the fishing bay below. There are gulls gathered on the shore, and walking down towards them is a fisherman with a bucket. He stands at the water's edge, backlit, and the gulls start to lift off and move in his direction. He steps back and swings the bucket's contents up and over the breaking waves – a thousand fishy fragments go flying, sparkling like pirate gold in the sunrise. The gulls, squalling excitedly, fall upon the booty, as the fisherman turns and walks back up the beach.

The word 'commensal' is a handy one when talking about gulls, and other animals which have a link to people. It means, literally, 'eating at the same table' or 'eating the same meal' but that doesn't really capture the sense of it as it's used in biology. Commensalism is used to describe a situation where one species depends on another for survival and is highly adapted to live alongside (or even actually on top of) the other, but the other party is not affected by the arrangement. So it is not the same as mutualism, where both parties benefit, or parasitism, where the dependent one harms the other one. Think of it more like a remora or 'suckerfish', which attaches itself to a shark's belly and spends its life being carried around and feeding on escaping fragments of the shark's lunch. The shark wouldn't care if it had remora passengers or not, but remoras have no other way to live.

Quite a few animal species are commensal with people. Well, compared to the great many animal species that have been comprehensively taken to the cleaners by people, it's a very low number, but it's still significant in and of itself. Some of them live very intimately with us indeed, such as the skin mite *Demodex folliculorum* (which lives in the hair follicles on our faces) and *Demodex brevis* (which also lives on our faces but prefers our sebaceous glands). Typically, neither of these little arachnids do us any harm, beyond making us feel a bit sick when we think about them munching away on our dead skin cells.

Then we have animals that live in our homes, if not actually on our person. These are creatures like German cockroaches, silverfish, bed bugs and various other unpopular insects. The brown rat and the house mouse are also pretty much commensal with people. Some of the animals that we have domesticated, over the centuries, probably started out as commensal with people, before the relationship became more mutualistic. It's likely that the wild ancestors of dogs and cats began to hang around human settlements and scavenge our cooking scraps, well before it occurred to us to train the former to do our bidding and make internet celebrities of the latter. The rats and the mice just sneaked in, exploiting a new and fast-growing food resource that allowed them to build up much bigger colonies than ever before. Thereafter, they travelled around the world with us, hitch-hiking in our wagons and stowing away on our ships, colonising every new land that we discovered and, often, quickly eradicating native island species that had no evolutionary grounding to cope with such efficient, omnivorous, agile and robust predators and competitors.

When it comes to birds, the domestication process that we suspect happened with cats and dogs may have happened in a similar way with domestic chickens and perhaps pigeons. Both of their wild ancestors – junglefowl and rock doves respectively – feed on the ground and would be drawn like magnets to spilled grain and general scraps. There are also various species of wild birds which have become highly commensal with humankind, though we've not had any good reason to go ahead and domesticate them, so they remain just our wild friends (and occasionally enemies). The house sparrow is a prime example. This enterprising seed eater is native to the Middle East but over the last few thousand years has spread throughout Eurasia as agriculture has spread. It's also been introduced to many other countries. Urban house sparrows nest in spaces in buildings, and mainly eat things that come their way via people. House sparrows can do a lot of damage in unsecured food stores, but are mostly pretty inoffensive (though they are problematic ecologically where they are non-native, such as North America, where they outcompete various native birds for food and nest sites). Not all house sparrow populations live alongside people, though, so this species can't be considered an 'obligate commensal'.

What about gulls? Herring gulls in particular, but some other kinds of gulls too in other parts of the world, are commensal with people in at least some places, and becoming more so, as coastal populations decline and urban populations increase. The scavenging side of their character, coupled with the careless and wasteful side of ours, fuels the relationship. Over most of the world, there's at least one animal species that has adapted to live in towns and has learned how to knock over and raid our rubbish bins. Where I live now, in inland Kent, it's foxes. Over much of the States, it's mostly raccoons, but sometimes brown bears as well. In parts of South Africa, baboons are bin raiders *par excellence* – but down by the seaside in Hastings it's herring gulls tearing the binbags to shreds.

Herring gull

The commensalism goes back further than the invention of the black rubbish sack, though. As soon as we worked out how to catch fish, the gulls wanted to join our team. On one of my coastal walks in south-west Cornwall, I paused on the scarily narrow cliffside track to look out at the immensity of the Atlantic, and saw a small fishing boat creeping along way out over the sparkly ultramarine water. In its wake it trailed a heavenly escort of gulls, wheeling around the stern with the light behind their wings. That boat's catch, hauled up from depths where it was out of reach to the gulls, drew the birds way out to sea, and then back again in to land to feast on scraps and leftovers. Back in the rough-hewn environs of Newlyn Harbour, the gulls look less heavenly and more mercenary, but they're there for the same reason – the chance of a free bite of fish. They loiter near the boats, and when the boats leave harbour, at least some of the hungry gulls will go with them. Gulls follow fishing boats, and it seems likely they have been doing this ever since people first made boats and nets, and took them out to sea.

Fishermen and fish-eating animals don't always get along famously. In the UK, people who own inland fisheries and angling ponds can legally kill cormorants and goosanders if they can demonstrate that these piscivorous birds threaten their livelihoods, and non-lethal control methods such as scaring have been tried and have failed. The level of vitriol some anglers have for cormorants in particular has to be seen (or read about on Facebook) to be believed. Some of these same wildlife-disliking freshwater anglers have also bemoaned the comeback of the otter in British waterways and called for a cull, spreading untruths about covert mass releases of captive-bred otters to try to stir up public feeling against them. This, unsurprisingly, isn't working – there can be few native British species more beloved than the otter. The anti-wildlife anglers seem not to mind kingfishers (yet) but certainly a minority of them look at grey herons somewhat askance, though plenty of other anglers love the waterline environment in all its diversity, and welcome all of the wildlife that shares their interests.

That's fishing for fun. By and large, sea fishing is a different kettle of … fish. Going out to sea to fish is mainly a commercial enterprise, and seabirds, traditionally, are quite popular with sea fishermen – watching for feeding seabirds can lead the way to a good spot to drop the nets, after all. Birds may follow boats, but boats might also follow birds. The most helpful seabirds in this respect are probably gannets, as they wander well offshore, and rather than being mainly scavengers and surface-pickers, will dive down and grab their own fish. It works both ways, too – a gannet knows that a fishing boat goes to where there are fish. A high-flying gannet can spot a boat from 11km away and will often approach to see what the boat has found.

This mutualistic relationship works with gulls as well. If you're a fisherman, a big swirl of gulls at sea is definitely worth checking out, and if you're a gull, a fishing boat is equally well worth a look. Gulls by and large stay near land, but like other seabirds they may follow actively feeding whales, dolphins and other large predators in the hope that small fish will be driven to the surface, or that the hunting animals will leave behind some scraps.

Switching this behaviour to follow fishing boats instead is not a great leap for the gull intellect and has proved much more profitable for them overall, because fishermen often chuck quite a bit of edible matter overboard. These 'discards' may be intact fish of the wrong species, or unfortunate young fish that are too small, or sometimes too many fish are caught and to satisfy quotas some of them have to be returned. Discarded fish may be alive and well and able to quickly escape to the depths, but quite often they are not alive, or at least not well, and are easy pickings for gulls. On land, at the ports and harbours where the caught fish are sorted and chopped up at processing plants, lots more edible waste is generated and this too is a valuable food resource for the local gulls, as long as it's left out in the open where they can get at it.

Even when they're not being directly helpful to the fishers, the gulls do no harm to speak of. They eat what's not wanted, and for fishermen spending hours or days at sea, their presence can be comforting. It's nice to have other living things around when you're out at sea for hours, especially if you are a lone fisher with a little boat, as many in the UK are. So the relationship between human and gull here, at least, is mostly cordial and sometimes downright affectionate – when the same gulls follow the same boats day after day, an attentive observer will soon learn how to tell them apart by their particular traits, and the birds' individual personalities begin to shine. I've read a few news stories about individual gulls forming a bond with fishermen, staying with the boats as they leave the harbour and return, and becoming as tame and friendly as a pet terrier – unfortunately sometimes also with a terrier-like feistiness and biteyness.

The UK fishing industry has, as we know, declined rather dramatically over the late 20th century and beyond. Let's look at some numbers. In 1938, there were nearly 48,000 commercial sea fishermen active in the UK, more than three-quarters of them working full-time at their trade. In 2017, the proportion of full-timers relative to part-timers was much the same (about 83% were/ are full-time), but the total number had plummeted to 11,692.

The steepest drop was between the late 1940s and 1960 when their numbers halved, and the decline has been steadily ongoing ever since.

Why this is happening isn't that simple an equation to solve. Part of the fall in personnel is down to improved efficiency of operations. And then there's a whole raft of issues around politics, quotas and ownership claims when it comes to which countries can take fish from the open sea that lies between them. But the core of the matter is that fish stocks have dwindled badly, especially in the North Sea, because of overfishing (and it's not just the fish we take but those we throw away too – high discards in some areas are undoubtedly a part of the problem). The tonnage of fish landed in Britain by our home fishing fleet in 1913 was 1.2 million, in 1960 it was still around the million mark, but by 2015 it was 0.415 million. And although there are now measures in place to help reverse the decline in our sea fishes, the impact on sea life around Britain, including gulls, has been considerable.

Here's one specific example. Canna is a little island in the Inner Hebrides, home to a couple of dozen humans and a variety of seabirds, including herring gulls, lesser black-backed gulls and great black-backs. Ornithologists have been running a bird count on the island for nearly 50 years, so have a very reliable and lengthy fund of data to examine. The numbers of each gull species nesting on Canna are not tremendous, but enough to provide a microcosm of the general UK gull trend over the last few decades. The count in 1988 revealed 1,525 pairs of herring gulls on Canna, plus 63 of lesser black-backs and 90 of great black-backs. By 2016, there were just 95 pairs of herring gulls, 13 of lesser black-backs and 18 of great black-backs. This dramatic fall coincides with plummeting fish landings at the mainland port of Mallaig, a 40km gull flight away. In the late 1980s and 1990s nearly 14,000 tonnes of fish were landed in Mallaig, on average, but this had fallen to an average of 4,456 tonnes per year between 2007 and 2014. As we've seen earlier in this book, all three of these large gull species have suffered significant population declines over the last few decades, throughout the UK.

Gulls and fishermen, it's clear, go together and rise and fall together. Fishing as an industry needs ongoing drastic action to save it, including expansive no-fish sanctuaries where fish can breed in peace, and limits placed on what we catch and when. Overfishing isn't the only reason behind falling fish stocks, though. Sea pollution is also an increasingly desperate problem that affects all marine life – fishermen have a vested interest in not throwing harmful rubbish into the marine environment, in theory, but studies have shown that discarded or lost fishing gear is the most significant component of oceanic plastic pollution. The tide of plastic pollution from land, plenty of which ends up in the sea, also needs radical action to solve. Climate change, too, affects the distribution and survival of fish.

For humans, food from the seas is of varying importance from place to place. The typical United States resident eats about 7kg of food that's of marine origin each year, but Japanese people eat, on average, five times as much as that. Worldwide, marine fisheries bring in about 140 million tons of living things each year, primarily fish, molluscs and crustaceans. About a quarter of this tonnage goes into animal feed in the form of fish meal and fish oil. Many different groups and species are used – among the most important 'food fish' are herring, mackerel, cod, tuna, pilchard, anchovies, pollock, capelin and haddock, plus a variety of flat (round) fish.

So, we humans are eating the seabirds' food, but it doesn't end there. Seabirds themselves, and their eggs, are also harvested for food. This is a big deal in some parts of the world – for example, in Tasmania at least 200,000 short-tailed shearwater chicks (aka 'muttonbirds') are taken from their burrows (under licence) each year, and up to 250,000 seabirds (mainly fulmars and puffins) are killed on the Faroe Islands each year for food.

This kind of thing doesn't happen very much around the British Isles any more, and is certainly not a current issue affecting our gull populations, but certain island communities were long sustained by their breeding seabirds. I had the good luck to visit St Kilda, way off north-west of the Outer Hebrides, a couple of summers ago. The main island, Hirta, supported a couple of hundred hardy humans over the last two millennia, the remnants of which (just 36 people) were evacuated to the mainland in 1930 after successive crop failures and other problems had made survival impossible. The last of them, Rachel Johnson, died in 2016 – at 93, surely a far greater age than she could have reached on a sporadic diet of seaweed and seabird. The islanders' stone houses now stand empty near the shore, except for those now converted to be accommodation for the wardens in summer, and one which is a museum. The island now belongs to the birds which the islanders used to catch and dry out as winter food supplies – the auks, shags, gannets and gulls.

Baby gannets ('gugas') are still harvested on the small, uninhabited island of Sula Sgeir, which lies somewhat near St Kilda. Only 10 people from the island of Lewis are licensed to collect them, and they take no more than 2,000 young, nearly fledged gannets for human consumption. The hunt is controversial, and also dangerous, though considered sustainable at least at the moment, as gannets are bucking the general UK seabird population trend and increasing at present.

Seabird eggs, including those of gulls, were a valuable source of protein for Scottish islanders in years gone by, and on the mainland, too, a great many eggs were traditionally collected from black-headed gull nests. These colonies had the advantage of being quite easily accessible, and gull eggs were and are considered to be particularly tasty. The adult birds themselves were also killed for food, but it was the egg trade that was particularly intense, and lucrative, as we saw in chapter 1. Today, eggs are still taken by a few licensed collectors, but at a much lower rate, and within a strict time frame. This means that the gulls' breeding success is not necessarily harmed by the harvest, as pairs have time to re-lay and successfully raise a brood of young, though for some pairs the energy expenditure in producing their first clutch will have been too much and they will not be able to have a second try.

Kittiwakes and their eggs were also formerly harvested for food, from before the 18th century. Their cliffside colonies were assailed by brave climbers armed with ropes and baskets. The birds' feathers were valued for hat decoration, their dense white body plumage used as a sort of fur, and some people even enjoyed shooting the birds merely for fun. These folks would charter a boat that would sail alongside the breeding cliffs, and when it got close the crew would sound a siren to panic the kittiwakes from their nests. The 'sport' of shooting the whirling birds out of the sky then began, and thousands of nesting birds were killed, leaving their chicks to die too. The colonies at Bempton Cliffs and Flamborough Head in Yorkshire were particularly heavily targeted. As the killing intensified, the falling numbers of nesting kittiwakes began to spark worries, not just for the birds themselves but also for mariners, who relied on the kittiwakes' raucous calls in heavy fog to warn them when they were close to land. So, in the 1860s, concerned parties began to lobby against the shooters, and in 1869 the Sea Birds Preservation Act was passed, giving some (though not full) protection to the birds during the breeding season. The RSPB itself was born out of pressure groups campaigning against the killing of kittiwakes (and other wild birds) for their feathers, in the late 1800s. By the 1950s, kittiwakes were largely being left unmolested at their nesting grounds. However, new and much more insidious threats were waiting in the wings to make their lives difficult.

'Our' kittiwake, the black-legged kittiwake that breeds on both sides of the Atlantic, is today considered 'Vulnerable' in a global conservation context by the International Union for the Conservation of Nature. Many other seabirds are in dire straits too, including other British species. The puffin is also 'Vulnerable', and its auk colleague the razorbill is 'Near Threatened'. Our sea ducks are rather beleaguered – the eider duck is 'Near Threatened' and the velvet scoter and the long-tailed duck, a pair of fine-looking sea ducks that winter off our east coast, are both 'Vulnerable'. Among the petrels and shearwaters, Leach's storm-petrel (which breeds on some remote Scottish islands) is another in the 'Vulnerable' category, while the Balearic shearwater, a regular migrant visitor to our waters, is classed as 'Critically Endangered', in other words facing an extremely high risk of extinction in the wild. That's the worst possible classification for a species to have while it still lives on in the wild. Keeping and breeding such long-distance seafarers as shearwaters in captivity really isn't feasible, so the Balearic shearwater is definitely looking down the barrel of total extinction very soon if we don't take drastic action to save it.

A review of published data between 1950 and 2010 revealed that seabird populations worldwide have fallen by almost 70%. This reflects what's going on under the surface – seabirds are simply the most easily observed element of the marine ecosystem, but their struggles are echoed in underseas environments around the planet. The British Isles supports internationally important numbers of breeding seabirds with just under eight million birds from 25 species nesting here. They include 90% of the world's Manx shearwaters, 68% of gannets and 60% of great skuas. Some of those 25 species – including gannets and great skuas – are doing pretty well here at the moment, with recent population increases. But others are struggling badly, including the kittiwake, and also puffins, and some terns – all birds that feed mainly on sand eels and other small fish. These species are probably suffering the consequences of climate change-driven declines in the plankton population on which small fish feed, and they are also affected by plastic pollution. The region of the mid-Atlantic ridge, a vital wintering area for kittiwakes halfway between us and North America, is steadily warming up, which is making it less hospitable for plankton and all the organisms that depend on plankton. Kittiwakes are struggling to make it through winter, and those that do make it are less likely to breed successfully. They continue to rely on small fish through the breeding season, though at this time of year they forage much closer to their breeding grounds. In years where sand eel harvesting in the North Sea is heaviest, kittiwake breeding failure increases.

Most seabirds are highly adapted to their particular way of life. The kittiwake and its ecological allies are closely dependent on particular elements of the marine environment, and so have no way to escape those problems with our seas today. Some other gull species are beset with similar troubles, or are restricted to very small geographical ranges, and are also considered at risk of extinction. Worldwide, four species are 'Near Threatened' – Olrog's gull, Heermann's gull, Armenian gull and ivory gull. Another four besides our kittiwake are 'Vulnerable', including its Pacific counterpart, the red-legged kittiwake, along with Saunders's gull, relict gull and lava gull. In world terms, the gull nearest to the edge of existence is the black-billed gull, the only gull species that is endemic to New Zealand. It occurs patchily over both islands but its numbers are concentrated at the south of South Island, and the rapid decline that propelled it into the 'Endangered' category has mainly been caused by non-native mammals eating its eggs and young.

(Before I leave the black-billed gull, I can't resist mentioning here the 'gullforglory' Instagram page. This page was created to drum up support for the black-billed gull as a 2017 contender for an annual New Zealand 'Bird of the Year' competition, run by the conservation organisation Forest & Bird. As well as including a photo of an adorable baby black-billed gull in a victorious pose, the gullforglory creator also posted many photos of other NZ birdlife – the competitors –with derisory captions. The black-fronted dotterel was branded a 'beach-breeding moron', the takahe a 'homeless chicken' and the kereru or New Zealand pigeon an 'overweight tree rat'. Sadly, the gull did not win – the honour fell to the kea, a charismatic mountain parrot, but the gullforglory page remains active today. What they said about the kiwi can't be repeated in a family-friendly book, but I'll leave the link here in the hope that it stays active as long as possible, and you can look it up for yourself: www.instagram.com/gullforglory.)

The other gulls in the world are all classed as 'Least Concern', and some are doing very well at present, with large, widespread and perhaps even increasing populations – in some cases, they are colonising new habitats and even new continents. These successful species have the advantage of adaptability – through their natural willingness to travel and to exploit new opportunities, they have found that there's more than one way to be a gull. This also means that there are more ways than one for gulls' lives to intersect with those of humans.

The history of life on Earth is full of similar stories. In early human history, several species around the world (some of which we've looked at the start of this chapter) took the opportunity to become commensal with humans and generally did very well out of it. In more recent times, adopting the habit of commensalism is often not so much a new opportunity as a last-ditch survival attempt, as human influence spreads ever wider and the natural world changes more and more rapidly. The three large gulls that breed widely in Britain – the herring gull, lesser black-back and great black-back – are all suffering severe declines in their traditional, natural habitat, and it could be that only the small proportion of the population that opts to live full-time alongside humans will prevail.

But what if the humans are not up for that? The trouble is that few commensal relationships will go entirely unnoticed by the species that's passive in the relationship. We humans notice when other animals live in our spaces and use resources that we want for ourselves, or even when they just use the stuff that accumulates around us that we don't really need. Our species has an inglorious history when it comes to sharing our lives with other animals. If those animals make themselves noticed, even just a little, we don't like it, and we punish those animals without mercy.

If you take a look at the shared history in Great Britain, for example, of humans and the native predators that live (or, in some cases, lived) here too, you'll find a long, bloody and sorry story. As soon as we started managing land to keep livestock and to encourage wild 'game', and bringing new kinds of animals over from abroad to shoot, conflict arose. The amount of wandering deer and estate-reared pheasants killed by native wild predators might have been very small but it was still utterly intolerable to the landowners, who saw no reason whatsoever not to kill every predator they could. The result of this was the extirpation of all our larger predators (the lynx in AD 400, the brown bear in about AD 1000 and the wolf by 1680). Smaller mammalian predators proved harder to eradicate and they all hung on, but several were brought to extremely low numbers – they include the pine marten, Scottish wildcat, polecat and otter. The logbooks of gamekeepers from big rural hunting estates tell of these mammals being wiped out in their hundreds and thousands, reducing these four once common and widespread species to tiny remnant populations, surviving only in the most remote and rural corners of Britain.

We also wiped out four species of predatory birds in the 19th and early 20th centuries – the goshawk, marsh harrier, white-tailed eagle and osprey. Happily, all have now returned to us, some with a bit of help through reintroduction projects. Nearly all other birds of prey suffered massive losses from this relentless persecution, only saved from the same fate as the extirpated foursome when new legal protections were introduced in the mid-20th century. Several of them then suffered a further blow when indiscriminate DDT use swept the country's farmland through the 1950s and 1960s. This toxin, used as an insecticide, accumulated in the food chain, killing many birds of prey and damaging and destroying the breeding capabilities of many more. Its persistence in animal bodies means that even today, more than 30 years after its use was banned in Britain, it can still be found in the tissues of birds of prey and other carnivorous animals.

DDT also affected and continues to affect gulls. In North America, where it was used just as enthusiastically as in the UK, it was found to cause eggshell thinning and feminisation of male western gull embryos in the egg – a lack of viable males, because of this DDT poisoning, could be the reason behind the high proportion of female-female pairs of western gulls that were noted in California in the late 1970s. And high levels of DDT contamination, along with other pollutants, is still affecting the lifespans and eggshell thicknesses of ivory gulls in the high Arctic.

Humans are not good at playing with others, and we're also not good at learning from our mistakes. Through hard work on many fronts, conservationists in Britain have restored (somewhat) our beleaguered bird of prey populations, but already there are calls in some quarters for culls to be carried out. Hen harriers eat a few red grouse chicks, buzzards might prey on pheasant poults, and peregrines and sparrowhawks catch a few racing pigeons. This is inevitable – any artificially inflated population of prey will attract predators. The law protects all wild birds of prey from harm, but unfortunately illegal persecution of raptors remains a serious problem – and an awfully high proportion of those crimes take place on game shooting estates. Research on satellite-tagged hen harriers, for example, shows that they are 10 times more likely to die or disappear on grouse moors than in other habitats. It's easy to imagine history repeating itself and all raptor populations dwindling dramatically again, if the calls for a right to cull raptors grow loud enough and the legal situation is changed.

When it comes to gulls living alongside humans, the problem is more one of daily inconvenience for everyone, rather than a minor impact on the leisure activities of a few. Having grown up surrounded by urban herring gulls, I know the nature of these inconveniences personally and spoke of some of them in chapter 3. To recap – your car and your patio and sometimes your person will be splattered with gull poo. Your sleep will be disturbed by gull shrieks from first light. You and your pets may be dive-bombed by angry, aggressive gulls in the breeding season. On bin day, if you put your rubbish bags out too early, they will be ripped up and their contents strewn across the street. Baby gulls will fall in your garden and live there for weeks (or die there) if not rescued and returned to the rooftops. Your beach picnics will be disrupted, and your garden bird table commandeered. But that's not the half of it.

If you want to know just how diabolical gulls really are, just start looking at some newspaper and news website headlines. Here's a few to be going on with: 'Seagull menace: Truth about Britain's new Public Enemy No1 following spate of attacks' (*Daily Mirror*). 'What happens when seagulls attack? "It's a war zone, we are even seeing people with mouth injuries"' (*The Independent*). 'Dive-bombing seagulls have trapped an OAP couple in their Kent home all summer (*The Sun*).

'Pet dog pecked to death by seagull in "horrific" garden attack' (*The Telegraph*).

That last story, from July 2015, is a particular shocker. The dog in question, a Yorkshire terrier, was attacked by a dive-bombing gull in the back garden of a house in Newquay, Cornwall, and pecked repeatedly on the head. The dog had to be put down because of the severity of the injuries inflicted by the gull's bill. A couple of months earlier, the *Telegraph* had reported of the same fate befalling a chihuahua in a garden in Honiton, Devon. Now, I have been pecked on the head by Arctic terns, as I mentioned earlier, when I was walking through their nesting area on Inner Farne island. Their blows were forceful and it was pretty damn painful, even though I knew it was coming and had a hat on, and an Arctic tern is much smaller and lighter than a herring gull, and I am much, much bigger than a Yorkshire terrier. So I have no trouble believing that a gull could do this, if riled up in defence of its nest and correctly recognising a dog, even a very little one, as a predator.

(The BBC news website had a similar story of another Cornish family losing a much-loved pet to a marauding gulls. This, too, occurred in 2015 but the pet in question this time was a tortoise, which was ambling around in the family's Liskeard back garden when the attack happened. The gulls flipped the reptile over to attack the softer underside, causing injuries which proved fatal. Although this, like the other attacks, took place during the breeding season, it's hard to see it as a nest defence situation, as tortoises are extremely non-threatening herbivores. So I can only guess that the gulls saw this poor tortoise as a plaything, or slow-moving food.)

What about those unfortunate OAPs who were trapped in their home all summer long? People, unlike tortoises, do present a real threat to gull nests, at least in theory, so gulls will attack people who get too close. This is an extreme case of something that happens rather a lot. The Smiths, both in their eighties, live in a bungalow in Walmer, near Deal in Kent, and in 2018 were joined by a pair of problem neighbours. The rooftop-nesting gulls would swoop relentlessly at the couple whenever they ventured into their back garden in daylight hours. Mrs Smith had to go out armed with an umbrella to fend off the birds when she watered her tomatoes. Pleasant outdoor lunches in the summer sun became impossible. Much as the wording of the headline made me laugh when I first saw it, I read on and soon felt very bad for the couple. Even though I'm pretty comfortable in the company of gulls, I too would have avoided the garden if subjected to this sort of treatment. It is rare for rooftop-nesting gulls to be *this* aggressive towards the people who live under said roof, but perhaps because this house was a bungalow, the nest just wasn't high enough for the gulls to feel safe about activity below.

My favourite of all the news stories I found – and I must admit I'm still not even sure it wasn't a joke and did check the date of publication (it wasn't April 1st) came from the website devononline.com. It dates from 2018 and reads '"Angry church seagull keeps attacking me and all I want to do is play Pokémon Go" says Devon man.' Reading on, I learned that the man in question was being targeted individually by a (presumably herring) gull as he tried to reach a particular Pokémon-hunting hotspot in St Peter's Churchyard, Tiverton. (For those who don't know, Pokémon Go is a mobile phone game that involves exploring the real world to find and capture virtual cartoon monsters. All kinds of fun.) Other churchyard visitors were ignored but this man was persistently dive-bombed and wondered if his bald head could be enraging the bird. The article was illustrated with a brilliant photo of the victim from a gull's-eye view, pulling an alarmed face and holding up his arms to fend off winged assailants, which to me suggested a certain sporting acceptance of the situation on his part. Another picture showed him batting away an Australian silver gull, a fine testament to the designer's Photoshop skills, if not to the article researcher's attention to detail.

The 2015 *Daily Mirror* story headlined 'Seagull menace: Truth about Britain's new Public Enemy No. 1 following spate of attacks' includes photos of a woman's bloodied head – cut open by a gull that attacked her as she and her dog were walking down a street in Helston, Cornwall (Cornwall again! What is it about the Cornish gull population?). Apparently, the gull was guarding its chick, which had fallen from a rooftop onto the pavement opposite. The article talks of townsfolk from Brighton to Bristol and right across Scotland calling for a gull cull, and ends with a poll asking 'Do you think there should be a cull on seagulls?' I had to cast my vote (a no, of course) to see the other results – 77% of more than 500 respondents had voted yes.

On to *The Independent* in 2017, and 'What happens when seagulls attack? "It's a war zone, we are even seeing people with mouth injuries"'. The war zone is, once again, Cornwall, and the mouth injuries are the result of incredibly bold herring gulls grabbing food right out of people's (often children's) mouths. The reporter spoke to a pharmacist who sees about two human victims of gull violence a week at her pharmacy, and suspects the true toll is much higher. These gulls, of course, aren't even being aggressive as such, just hungry. They don't see the child's delicate face, they only see the chip that's going into it.

This article mentioned a Facebook page called 'We need a Seagull Cull'. I went off to look for it but didn't find it. Could attitudes be softening? I did find a group called 'Support a Seagull Cull', with the robust description 'Vile killing machines of the sky', but it had only 20 members and had been inactive since 2015. I did, however, find two pages emphatically arguing the opposite point – 'We do NOT need a cull on gulls', and 'We DON'T need a Gull Cull'. Even the group 'Fife Council. Fix the Seagull Problem' doesn't argue for a cull as such, just for businesses and individuals to take non-lethal steps to discourage the birds from nesting in the town (and it seems to recognise that this would mean providing safer nesting places along the wild coast).

I also found some news stories recounting acts of deliberate cruelty towards gulls. In Britain we have some strong laws against cruelty to animals, including wild animals, as well as the Wildlife and Countryside Act that protects all wild birds. So it's good to see that the perpetrators in some of these stories were indeed charged and sentenced appropriately. But the RSPCA still takes many calls every year from concerned members of the public who've witnessed people shooting at gulls with air rifles, illegally destroying their nests, trying to poison them or run them over, and so on.

I've talked to a lot of people who really can't stand gulls, and they've all taken pains to clarify that they would not ever personally hurt a gull, even if they'd be supportive of an official cull. Most of them had no idea that gulls are protected by law, either, so it's not fear of legal consequences that is stopping them. The sort of person who'd kick a gull to death might also do the same to a cat or a dog or any other animal that got in their way, and thankfully such people are rare. However, it wouldn't do any harm to get the message across much more loudly and clearly that it is illegal to harm a gull or its nest, even in a humane manner, unless you have the correct licence (including meeting the conditions of the General Licence in the case of the lesser black-backed gull and the nests of herring gulls – but see footnote on page 96). Prosecutions can and have happened. For example, in 2017 two men were fined £1,000 and £500 for killing a gull (and also an oystercatcher) with a slingshot in Scotland. In 2019 a man in Somerset was sentenced to a 12-week curfew as well as paying hefty court costs for killing a lesser black-backed gull by grabbing it and swinging it against a wall. Several people witnessed the attack and their reports led to the man's arrest. A Wirral man narrowly escaped prison after being found guilty of shooting a gull on his roof in 2015, instead receiving a fine of nearly £1,000. He claimed to have had no idea of the bird's protected status. Perhaps it would have made no difference, but successful prosecutions like this help raise awareness and push us to do the right thing. An RSPB officer involved in yet another case (in this case, a Cornwall teenager charged with shooting a pair of nesting gulls with an air rifle) summed it up thus: 'Gulls have a particularly tough time of it and every year we see callous attacks like this, particularly in coastal areas. We urge people to be tolerant of the wildlife living around them, and remember they are protected under law.'

So, birdwatchers and fishermen notwithstanding, the fact is that a lot of people aren't keen on gulls, and a handful of people would even actively murder them, despite legal protection. I have to end this chapter on a high note, though, and celebrate the views and activities of some of the ordinary people out there who just happen to be fond of gulls. Let's begin with a team of academics at the University of Bath, who carried out a public engagement project in their city in 2017. Bath is one of those places that has a sizeable and growing roof-nesting gull population despite not being on the coast, and in general its people aren't impressed with this situation. The project centred around educating children about the problems herring and lesser black-backed gulls face in their natural coastal breeding habitat, which is driving the tendency to build inland populations. The young participants also made gull puppets and filmed scenes of them interacting and storytelling. The children's attitudes towards gulls had become considerably more friendly by the project's end.

People who work in wildlife rescue and rehabilitation centres tend to get to know gulls a lot better than most of us. At the bird hospital at Mousehole, Cornwall, there are several long-term resident gulls living together in comfortable accommodation. These birds suffered injuries that caused permanent disability, making them unsuitable for release, but appear to be quite happy in their aviary, with access to a pool and as much tasty food as they want – and they are much loved by the staff.

Urban gulls are particularly susceptible to misadventure, thanks to their curious and enterprising ways. In 2016, the Vale Wildlife Hospital near Tewkesbury took in a particularly adventurous gull that had fallen into a vat of curry at a food factory, leaving it almost unrecognisable. The bird, entirely orange and (by all accounts) smelling delicious, was caught by staff at the factory and taken to the hospital where its plumage was cleaned up, revealing a handsome if sheepish adult herring gull who was returned to the wild in short order. And in 2018, according to an article in *The Independent*, there was an outbreak of alcoholism among the gulls of south-west England, with more than a dozen taken in by the RSPCA after being discovered wobbling along on various beaches in Devon and Dorset, disorientated and reeking of booze. Most recovered well after being sick. The source of the problem was not located, though the RSPCA appealed to local breweries and other alcohol producers in the area to check that their waste was not gull-accessible.

Perhaps these news stories can help us warm to gulls somewhat and discover kinship with them – a liking for curry and beer is, after all, a trait they share with many of us. The internet tells me that plenty of people keen on mysticism and suchlike have selected the 'seagull' as their spirit animal, rather than something more glamorous like a wolf or eagle. Having a seagull as one's spirit animal means you are charming and fun-loving, light-hearted, sociable and with a magnetic personality. (No doubt it also means you steal chips, but hopefully you draw the line at biting children's faces and pooing on parked cars.)

Some other people have advanced their gull identity one stage further, thanks to a rather marvellous new gull-related event that took place in the Belgian town of Adinkerke in 2019. This was the inaugural European Gull Screeching Championship. As the name suggests, this contest challenges people to do their best to make gull noises. In the Adinkerke region, as with many British towns, gulls are viewed none too kindly by a large swathe of the general public, because they do all the usual gull things. The championship was devised by some local gull fans as a way to foster warmer feelings towards the birds. Contestants came from as far afield as the Netherlands, France and Belgium; several of them donned gull costumes of feathery hats, googly eyes, bills and wings for their performances to enhance the impression of gullishness. The video clip I watched of them doing their stuff was really very impressive. Most chose to mimic the yodelling 'long-call' of the herring gull, and two in particular were quite outstanding, walking away with a shared victor's title. I'm not sure how many entrants there were in all but from the video it looked like a few dozen at least – hopefully next year there will be many more. Event judge Jan Seys, from the Flanders Marine Institute, pointed out that the competition involved 'real science', as only those who've spent time watching and listening to gulls in the wild would be able to produce a really good gull impression.

Whether it's love, hate or hilarity, gulls spark strong feelings in us humans – they always have. We have a long shared history, of conflict as well as cooperation. Today, though, they need our compassion above all else – for we can no longer deny the impact we are having on the natural lives and survival of gulls, nor the fact that our own cities offer the best chance of their continued survival.

CHAPTER 5

Gulls in Word and Image

Kittiwakes

'MINE?' SAYS THE gull, hopefully. Another glances over. Some more long-necked, frighteningly blank-faced and beady-eyed gulls flap down and settle nearby, then more. Soon a large flock has assembled. They tilt their heads as they scrutinise their quarry – Dory and Marlin, the lead characters in *Finding Nemo* – flapping about on a wooden pontoon. The two fish leap into the capacious bill of their unlikely friend, Nigel the pelican. Nigel takes flight, the gulls do the same, and a short but wild chase scene ensues. 'Mineminemineminemine' chorus the gulls as they pursue the pelican around the harbour, until he tricks them into flying headlong into an unfurled sail.

Our 20th and 21st century fictional gulls are often either quite evil or just not very bright, and the *Finding Nemo* gulls, with their limited vocabulary and the ease with which they can be tricked, fall firmly into the latter category. Their overriding motivation is to fill their tummies, and this harks back to their very name – *Larus* and by extension Laridae comes from the Greek word *laros*, meaning 'a ravenous seabird'. The word 'gull', incidentally, has a Celtic root (Welsh *gwylan* or Breton *gwelan*, meaning 'wailing').

Did the gulls of older folk tales and mythologies have any redeeming traits? Some of them certainly did. While researching gull-related folklore, I came upon a particularly charming story – that of St Kenneth of Wales. This 6th century saint was, it's said, placed in some kind of floating cradle when he was born and dropped, Moses-like, into a stream. His parents, or the community at large, did this to dispose of him because he was a cripple. However, they reckoned without the local black-headed gulls, who took pity on the baby and rescued him, carrying him up to their clifftop nesting grounds on the Gower Peninsula. Here they made him a comfy bed from their own feathers, and fed him on doe's milk (exactly how they managed to convince the doe to provide her milk is, sadly, not known). I liked the idea of a saint raised by gulls. He lived on the Gower as a hermit in his adult life (perhaps because he could only communicate by shrieking?) though according to some sources he did become friendly enough with a human female to have a child with her. At some point later in his life, he met St David, who took it upon himself to cure St Kenneth's bad leg. Kenneth was outraged by this unsolicited divine medical assistance and demanded that the leg be restored to its former state.

Another saint with a link to Laridae was St Bartholomew, who was born in the 1100s in Northumberland. His ancestry was Scandinavian and his given name was Tostig, but he felt the calling of the monastery at a young age. He was a monk at Durham (under the name William) for some years, but then after seeing St Cuthbert in a vision, he decided to become Bartholomew instead and to live out his days on Inner Farne in the Farne Islands, where St Cuthbert had lived as a hermit some 400 years before. St Bartholomew lived in St Cuthbert's old cell for 42 years on Inner Farne. Anyone who visits Inner Farne today will know that any time spent on this island means intimate acquaintance with a great many seabirds. Most of the Farnes seabirds are only there to breed, between April and August, but gulls linger year-round and perhaps this is why St B's particular favourite was a gull (species unknown) which he tamed to the point that it would eat from his hand. (You're probably not that impressed by this if you've ever lost a chip and nearly lost the hand holding said chip to a hungry gull in a seaside resort.) One day a bird of prey killed this pet gull. Angered, St B captured the raptor responsible and kept it prisoner for a short time, but then realised that its predatory behaviour was natural, as God intended, and let it go again.

Native American folklore is a rich source of animal tales, and once you've read a few you will see that the same animals often have the same characters throughout – the trickster coyote and raven, the fierce and wise wolf, the powerful, straight-laced bear and the brave eagle. Gulls in native American mythology are usually rather base, rude and self-serving but I found a story of one which was both inventive and heroic. The story, told by the Innu, begins with a pair of bear cubs who wandered away from their den in search of tasty berries. While they were gone, a huge and evil monster killed and ate their mother, and then came hunting for the cubs. The cubs ran to the home of Grandma Gull, by the river, and pleaded for her help. Grandma Gull hid the cubs. Then she intercepted the monster and told it that the bear cubs had crossed the river, but she would give the monster a lift across in her boat. The monster climbed aboard but found the boat's fishy smell unbearable. When it dangled its head over the edge to avoid the smell, Grandma Gull produced a knife and cut the monster's head off. The two cubs lived safely with her by the river forever afterwards. (I'm struck, at this point, by what sorts of headlines might fill our national newspapers if gulls could indeed carry and wield bladed weapons...)

Another tale of gulls comes from the Cree, who didn't look too kindly upon them. Some settlers made their home next to a river, and made their living catching fish. Their activities caught the attention of the local gulls and also the local grey jays ('whiskey-jacks') which lived nearby. The jays came to pick scraps that were left behind when the people prepared the fish they'd caught, and always took what was offered in a quiet and polite manner. The gulls also came for scraps but their behaviour and attitude was aggressive and angry – as they snatched the scraps they shouted that all the river's fish rightfully belonged to them. Although the people thought the jays and gulls equally beautiful, over time they grew fond of the jays and disdainful of the gulls, and so today the gulls can only have whatever scraps they find floating on the water, while the jays are allowed into fishing camps to help themselves. It's clear that gulls everywhere have failed to learn this lesson.

Gulls are often pitted against corvids in native American folklore. One Alaskan story started out with Gull and Raven on friendly terms, so much so that Raven invited Gull to the campfire to share a meal of cooked fish. Gull turned up with many other gulls in tow – the flock ate all of the fish, leaving nothing for Raven. Their greed cost them dearly, because Raven cast them into the fire as punishment. The flames burned their wingtips before they managed to get away, resulting in the black wing markings that many species show. Other similar stories cast the gulls as unintelligent as well as impolite, and have them plunging into the fire of their own accord to steal food as it cooks.

The gull in a Kwakiutl story had a pivotal (if passive) role. Raven, the creator of the people, decided to provide them with fish to catch, and did this by jumping on Gull's belly to make him regurgitate the first fish into the river. Another gull versus raven story comes to us from the Chinook. Here, Gull and Raven are far from friends and are at war, each leading an army (of seabirds and land birds respectively). Although Gull killed Raven, Raven's sister Crow took over the role of army general and the land birds prevailed in the end. As her spoils of war, Crow demanded strandline foraging rights, and this is why you'll see gulls and corvids feeding together on the beaches at dawn.

The Welsh Celts believed that gulls could predict sea storms – their presence inland was a sign that heavy weather was rolling in. Their calls to each other when heading landward meant: 'Heavy weather, heavy weather, let's go to the heather,' and when they returned to the sea they cried: 'The storm is no more, let's go to the shore.' An Irish and Scottish chant along the same lines goes like this: 'Sea-gull, sea-gull, sit on the strand, it's never good weather when you're inland.' The calls of the gulls foretold actual disaster at sea.

Reincarnation is a common theme in folk tales, with dead humans reborn as appropriate animals (though almost never the other way round, curiously). Gulls in many coastal nations' mythologies are vessels for the spirits of lost sailors, fishermen and anyone else who drowned at sea. These spirits are often considered malevolent. Their gaze should be avoided, lest they come and peck out your eye, and you should certainly not consider killing one. A trio of gulls flying together was a particularly bad omen. However, in other accounts the gulls are only temporary vessels, to carry the seamen's ghosts to rest. If the ghosts are angry and vengeful, they will not be carried away by gulls but will roam around unbodied, whipping up sea storms and the like.

A Scottish story tells that a pair of gulls calling to warn of a storm are the spirits of two star-crossed lovers – one the daughter of a Pictish king and the other a Scots warrior. They fled to the island of Islay to escape the war between their peoples, but their boat was destroyed on the way by a sea storm and they were lost to the waves. Now their spirits live on as gulls, yelling warnings to others of storms on the way.

If you take a look at the USA's various official 'state birds', you'll see the same names come up again and again. Seven states have opted for the northern cardinal, a cute songbird with a jaunty crest and (in the male) bright scarlet plumage. The western meadowlark, another sweet-singing and colourful little bird, is picked by six states. Most of the rest are also songbirds, with a few bald eagles and a smattering of gamebirds like quails and turkeys for the states with a particular fondness for huntin' and shootin'. So, in general, state bird choice rationale is along the lines of 'it lives in our state and looks pretty/sings nicely/is fearsome/tastes good'. The less impressive birds, for want of a better word, don't usually get a look in. One that stands out like a sore thumb is Utah's state bird. This state has opted for the California gull as their avian symbol. While California gulls are not particularly noted for being beautiful, tuneful, majestic or delicious, what they did for Utah goes way beyond such considerations.

The 'miracle of the gulls' took place in 1848, so it's told, when some 4,000 Mormon pioneers were settling in the Salt Lake Valley. They arrived in spring 1847 and planted their first crops, making a small harvest from which they gathered seed for the next year. However, the following summer a vast plague of grasshopper-like insects (Mormon crickets) appeared and began to rip through the growing crops with devastating speed. The day was saved when flocks of California gulls arrived and feasted on the insects. The story goes that they ate all they could, then drank water and vomited before resuming the eating, so were clearly at work to save the crops from the insect pests rather than just satisfying their appetites – this is why the Latter-Day Saints consider the event a miracle, though no doubt it's become distorted (perhaps a lot) in the telling. Today, a handsome Seagull Monument stands in Salt Lake City in front of the Assembly Hall. This monument, a pillar at the centre of a round fountain, its round top surmounted by a pair of gulls cast in bronze, commemorates the event and honours the gulls.

The gull in Chekhov's 1895 play *The Seagull* is a symbol, but also a real (albeit dead) gull. In the play, Konstantin (a playwright) falls in love with Nina (a girl with aspirations to be an actress). He gives her a role in his new, experimental play, but it's a box office disaster. Later on, he makes a further attempt to woo her by bringing her a dead gull that he has just shot. It's a gift of love, but brings her nothing but confusion and distress. His gesture inspires Trigorin (a writer) to write a story, which he describes thus: 'A young girl lives all her life on the shore of a lake. She loves the lake, like a seagull, and she's happy and free, like a seagull. But a man arrives by chance, and when he sees her, he destroys her, out of sheer boredom. Like this seagull.' After talking with Trigorin, Nina falls in love with him but he rebuffs her. She runs away to Moscow to be an actress, but doesn't find any great success. Eventually she does get together with Trigorin, but he abandons her (after getting her pregnant) not long afterwards to return to his former lover (who just happens to be Konstantin's mum). Looking back on her life, she compares herself to the dead gull, though reasons that the comparison isn't valid because her choices were her own. Konstantin, who's pined for her throughout, tries to woo her one last time, fails, and concludes the sorry story by shooting himself.

Disappointingly, Chekhov doesn't provide us with any clues as to which species of gull his story features. That's not the case in *Watership Down*, the hugely popular rabbit-based saga written by Richard Adams in 1972. Whether this is a story for children or not is debatable – it might be about fluffy talking bunnies off on an adventure, but it certainly pulls no punches in the nature-red-in-tooth-and-claw stuff. Biological accuracy being a strength of Mr Adams's work, it's not surprising that the gull in *Watership Down* is of a named and real species. Kehaar the black-headed gull was found by the rabbits in a sorry state – alone, close to starvation, with an injured wing that rendered him flightless. Keen to recruit an avian ally who could scout out terrain for them, the rabbits take care of him and nurse him back to health. Later on, Kehaar helps the rabbits (who are all male) to find some females at another warren, so they can establish a new warren. A horrific bloody war results from this, but it's not Kehaar's fault.

In the book, Kehaar's dialogue is Norwegian-accented, resulting in much comedy (for example, his name for the rabbit Bigwig is 'Pigvig') though in the 1978 animated film adaptation, he sounds more Germanic (in my opinion). Either way, he is quite a rude and grumpy bird, though a loyal friend when the chips are down. He provides some much-needed comic relief in the film particularly. Although it's not as gruesome as the book, this film is a good deal more dark and violent than the average animal animation, and is rather infamous for traumatising children of my generation.

In 2018 a new serialised version of *Watership Down* was broadcast on the BBC. With an all-star voice cast, fewer scenes of bloody violence, and some nicely rendered (if somewhat hare-like) computer-animated bunnies, it drew in the viewers, though perhaps lacked the charm of the original film. In this version, Kehaar has become Scottish, which is perhaps more logical for a breeding-plumaged adult black-headed gull in England, but he is no longer quite the rude, hyperactive and feisty bird that we know and love, which is a pity. Also – and this is me being a massive pedant and probably won't spoil things for most viewers – he's no longer an adult black-headed gull but a first-summer, with retained juvenile tail and wing feathers. Also his vocalisations (before he speaks) are the gurgles and cackles of a Larus gull. (I want to say a herring gull but I am not certain – he's definitely not shrieking like a *Chroicocephalus* though.)

The gull in the book *Jonathan Livingston Seagull* is most definitely an unidentified gull – in fact the outline of the bird on the cover I'm looking at looks much more like a raptor, maybe an osprey. Of course, the author, Richard Bach, wasn't really concerned with larophiles' sensibilities when he wrote the book in 1970 (hence his use of the dreaded 'S' word). This is a book about flight, on the face of it, and its deeper themes explore love and forgiveness, self-betterment and spiritual transcendence. The titular seagull is uninspired by his fellow gulls' preoccupations with fighting over food and so on, and spends all his time honing his flight skills, learning wild new aerial moves, and eventually leaving Earth altogether and reaching a higher plane of existence where he meets and flies with other gulls like himself. He returns to Earth and becomes a flight teacher for the more adventurous young gulls of his original flock.

Richard Bach is himself a keen aviator, having done service for the United States Navy Reserve, the New Jersey Air National Guard and the 141st Fighter Squadron. In 2012 he was involved in a serious plane accident and that experience inspired him to write an addendum to *Jonathan Livingston Seagull*. This unpublished fourth section of the book, set many years later, describes how the long-gone Jonathan had become legendary in his flock, but that the gulls now spent more time honouring him in arcane rituals than actually practising the skills he had given them. One disenchanted member of the flock, one Anthony Gull, decides life is pointless and that he will end it all by plunging at breakneck speed into the sea. A reincarnated or ghostly Jonathan appears in the nick of time to save Anthony and help put things right again in the flock.

Jonathan Livingston Seagull seems to be loved and hated in equal measure, with some inspired by the layered meanings they find in its message and others considering it naïve and banal. The 1973 film adaptation, which was more or less 120 minutes of footage of gulls flying about, with spoken dialogue over the top, was not really loved at all, except for its Neil Diamond soundtrack. As well as enduring a savaging from critics, the film's director Hall Bartlett also had to contend with three lawsuits related to the film, including one from Richard Bach himself who was unhappy about various changes made to the storyline – a violation of his contract. As a result of this lawsuit, the film studio was forced to make some rewrites before the film could be released.

You won't find many other books for adults that have a lot to say about gulls. But passing mentions are abundant, and the same goes for poetry. I have a deep affection for the poem *In Country Sleep*, by Dylan Thomas, first published in 1947. In this lengthy work, Thomas goes full-on lyrical flight of fancy, with countryside-themed and often rather dark imagery stuffed into each stanza. In the second half we turn briefly from farm, forest and field to the seaside and find 'the lulled black-backed gull, on the wave with sand in its eyes'. His better-known *And Death Shall Have No Dominion*, about the dead finding peace (sort of) also speaks of Laridae – 'No more may gulls cry at their ears'.

One of my favourite children's books – *The Jungle Book* by Rudyard Kipling – includes a short story called *The White Seal*, set in Arctic Russia. It's about Kotick, a harp seal with a white coat, who goes on an adventure to find a new, safe place where his colony can live safely out of the reach of the sealers. It's full of Arctic wildlife, though sometimes Mr Kipling gives them names that weren't familiar to me. He explains that a 'gooveroosky' is a kittiwake and an 'epatka' is a puffin, and that a 'chicky' is a Burgomaster gull. I thought that a name like 'chicky' must denote one of the smaller, cuter gulls but I was mistaken – it's a glaucous gull, that big, beautiful white-winged bruiser of the far north. One Burgomaster gull has a brief but key role in the story, helping Kotick to find Sea Cow, who would help him on his quest. The gull explains that Kotick will recognise the possibly mythical Sea Cow (a Steller's sea cow, *Hydromalis gigas* – a relative of the manatees) because he is the only thing in the sea uglier than a walrus. *The Jungle Book* was first published in 1894, and the Arctic has changed somewhat since then, but even in Kipling's time the ugly old Sea Cow was no more – *H. gigas* was wiped out by overhunting by 1768, just 27 years after it was first discovered by Europeans.

In Christmas 2018 I discovered a brand new favourite children's story, when I unwrapped my presents and found among them the amazing picture-book *Steven Seagull: Action Hero* by Elys Dolan (2016). His name a cunning play on Steven Seagal, the hero of this story is so terrific that I don't even mind that I can't tell what species he's supposed to be in the illustrations. He's an ex-police officer who is called out of retirement to solve one last crime, alongside his faithful friend Mac, the goldfish (complete with bowl). The twosome has to work out who's been digging mysterious holes in the sand in Beach City. I won't spoil the ending but urge you to give it a read, even if you are not (strictly speaking) a child.

Another picture-book in which a gull takes the starring role is *Sidney Seagull Starts the Stomp*. This gull wasn't like the others – he wasn't any good at catching fish. But he learned how to charm worms to the soil's surface by stomping on the ground to imitate rainfall, and now all the gulls do the same thing. This 2014 book is by Emma Pickles and published by the charmingly named Gull and Buoy Publishing, along with various other seaside-themed picture books. There are quite a few other books of this genre out there in which gulls take the lead role, and some of them carry a serious message. *Sid the Seagull* by Marie Tonkin is one such – published in 2018, it tells of a friendly gull who gets all his food from kind people, but can't understand why the humans are becoming scared of him. In the story, Sid learns that for some people a friendly gull that gets too close can be frightening, and that he should look for his own food. Another tale, *The Story of a Seagull and the Cat Who Taught Her to Fly*, by Luis Sepúlveda and Satoshi Kitamura, tells of a young gull who was cared for by a friendly cat after her mother succumbed to an oil spill. As the most visible of our seabirds, and bristling with feisty characterfulness to boot, gulls frequently take a guest role in children's stories that are anything to do with the sea or shore.

From the adorable to the abominable – gulls are key antagonists in Daphne du Maurier's 1952 short horror story *The Birds*, and the 1963 Alfred Hitchcock film based on (or perhaps 'inspired by' would be better) that story. The original story is set in Cornwall and tells the tale of Nat, a retired war veteran, now working as a farmhand, who is in the thick of the action when the local birds go gradually berserk and start murdering the locals. Ms du Maurier knew her birds and name checks the like of sanderlings, redshanks and curlews, blue tits and bramblings, early in the story when a change in the weather seems to be causing them to behave oddly. Her gulls are gulls, too, not seagulls, and later we even get their full names – the black-headed and the herring gull, and the black-backed too (great or lesser not specified, but you can't have everything). The story is truly chilling, especially when the gulls gather at sea in such numbers that our hero mistakes them for whitecaps on the waves at first. As the tide rises, the gulls rise too and fly inland, joining flocks of crows and the like, and the horror begins as the suicidally violent birds attack people in the open, force their way into houses, and even bring down aircraft. By the end of the story, the military had failed to stop the birds, and all of Nat's neighbours have been killed. Nat and his family are trapped in their home and he is plotting how to keep the birds out over the days to come. The story reflects the horror of the Blitz in 1940 and 1941, and we never truly understand what has driven the wild birds to this action.

The film is different in almost every detail to the story – its setting and characters are entirely unrelated. However, the birds remain as soullessly violent as before and, once again, gulls are key to the action. The first attack occurs when the heroine is piloting her boat around Bodega Bay, California, to meet her putative boyfriend – a gull bounces off her face as she nears the shore. Soon afterwards, a children's birthday party is besieged by large numbers of gulls, and things rapidly turn to absolute chaos soon thereafter. The film is regarded with affection by many, and some of the scenes with the birds do stand up to the test of time (in others, the special effects don't look that special any more but that's only to be expected). A 1994 straight-to-TV sequel, *The Birds II: Land's End* was universally slated – the little I've seen of it involved mostly angry pigeons, which just aren't that scary.

When I watched the original *The Birds* film as a little girl, I couldn't bring myself to be scared of the birds. Indeed, I was very excited at the sight of all those tame crows (I longed for a pet crow myself). The only birds that were being obviously violent towards the people were the rather unconvincing model ones – the footage of real birds just created the impression of violence through tricksy camera work and dubbed-on sounds of shrieking. The birds in the film were mostly wild caught, but were tamed and trained by Ray Berwick, an animal wrangler of long pedigree and considerable renown. The American Society for the Prevention of Cruelty to Animals monitored activity throughout filming, and there was even a bird hospital on the set to treat any that were injured. A fair number of human cast members needed treatment too, for injuries sustained when birds went too enthusiastically for the meat baits that were provided to direct their movements. The scene in the book in which a gannet plunge-dives full speed at the hero's skull is not repeated in the film – I guess even the most inventive animal handler has his limits.

I'm not a huge fan of 1980s new wave and synth pop but those who are will speak in glowing terms of A Flock of Seagulls, a Liverpool-born band of this genre. Although their songs were not particularly gull-focused, the video they made for the song 'The more you live, the more you love', is filmed on the Giant's Causeway in Ireland and features lots of nice footage of swirling gull flocks (plus a rogue jackdaw). There is, or was, also a rather more obscure folk band called The Kittiwakes. This three-piece British group took their name and much of the inspiration for their 2009 album, *Lofoten Calling*, from a visit to the Lofoten Islands in arctic Norway, where they would undoubtedly have encountered many kittiwakes. The album cover artwork shows a pair of kittiwakes perched in a bare tree, growing out of a tiny island. The tree has impaled a canoe, and a third kittiwake rests on said canoe while a thoughtful lynx sits at the tree's foot.

Lyrics about non-specific gulls can be found on occasion in popular songs, including some of my personal favourites. In 2001, The Shins released their first single from *Oh, Inverted World*, their debut album – and this doleful track, 'New Slang', begins its chorus with 'And if you took to me like a gull takes to the wind'. Eagle Eye Cherry's album *Desireless* has a sea-themed track called 'When Mermaids Cry', which goes 'So, next time that the seagulls fly, don't you cry, sweet Lorelei'. And that lovely, lonely introverts' anthem, 'Wonderful Life' by Black (1986), speaks of 'gulls in the sky'. The internet reveals many more gull and seagull lyrics from a diverse span of songs, genres and years, and most of the mentions are highly poetic in nature – it seems that lots of musicians are inspired by gulls' grace, beauty and mournful voices.

An exception can be found on the 2010 album *My best friend is you*, by Britpop singer-songwriter Kate Nash. This work includes a track called 'I hate seagulls'. The opening line of said song is 'I hate seagulls'. I wanted to know why, but she doesn't say – the lyrics of the first verse just go on to list many other things that Kate hates.

Happily, that bit of throwaway gull contempt is balanced by the bizarre but strangely delightful 'Lesbian Seagull', a mellow number originally recorded by Tom Wilson Weinberg on his album *The Gay Name Game* in 1979. It gained wider acclaim when it appeared on the soundtrack to the 1996 animated film *Beavis and Butt-head do America*. In the film, the song is performed by a hippy, guitar-playing schoolteacher, to Beavis, Butt-head and their classmates, moments before he is arrested by agents from the Bureau of Alcohol, Tobacco and Firearms. It then reappears in the closing credits, this time sung by Engelbert Humperdinck, no less. Humperdinck's version was released in 1996 as a single B-side to the Red Hot Chili Peppers' cover of 'Love Rollercoaster', which also featured in the film. The song was inspired by ornithologists' widely reported discovery of a high proportion of female–female pairs at western gull colony in California (where else?) in the 1970s (see Chapter 2 for a bit more on this). Its lyrics describe the peaceful life of a lesbian seagull and her girlfriend, playing among the waves and sleeping in the dunes.

Gulls in myth and legend, lyric, literature and film, run the gamut of our complicated feelings about them. They're dangerous, greedy, violent, portentous, irritating, foolish, wise, helpful, beautiful, curious and talented. Just like real gulls ... well, sort of. Improving the public image of gulls can be carried out through storytelling as well as by fact finding, but there needs to be an awful lot of it to convince the hardcore gull-hating seaside town inhabitants that gulls deserve our love and respect.

CHAPTER 6

Larophilia

Caspian and herring gulls

R SPB RAINHAM MARSHES is a little oasis on the shore of the
tidal Thames. The nature reserve itself is reborn wild marshland,
once MOD terrain but restored to something like its original beauty
after years of hard landscaping work, but the gravelly banks of the river
are also well worth a look. I like to visit on sunny days in early autumn
to take photos of dragonflies, and if the forecast is to be believed,
today is going to be a beauty. So I've taken the train here. It's a short
walk now, from Purfleet station along the Thames path. I arrive just
as the place opens and one of the volunteers is standing by the main
door into the visitor centre, aiming a telescope at the far riverside.
He excitedly invites me to take a look, so I do. Through the scope,
the circle of magnified shoreline is full of moulting large gulls, a big
flock of them loafing sleepily near the water's edge and shimmering
gently in the building haze. The one in the centre is side-on, showing
a little dark eye in a big white head, and a long parallel-edged bill. The
volunteer's already telling me it's a Caspian gull, a visitor from western
Asia. I might have worked out what it was for myself, because I know
that 'caspo' look from a few previous encounters with the species, but
that's by the by. There's no way I would have found it in the first place.

All gulls, wherever they live on Earth, are cousins to each other and all are recognisable as gulls to anyone who cares to take a good look at them – you don't need to be any kind of bird expert to know a gull when you see one. But which gull is it? Most birders will wince at you if you dare to point out a flock of 'seagulls'. If they are the stroppy sort of birder, they might also tell you, through gritted teeth, that there are more than 50 species of *gulls* in the world, and actually, in fact there are four different species in that flock alone

There is another kind of birder who will behave a bit differently though. This one, when you point out the flock of gulls, won't say anything much, but will set up his or her (but I have to admit that it's overwhelmingly likely that this person is a man) telescope and begin scanning slowly through the gulls, one at a time, making notes as to species, subspecies, sex and age class for each individual. And when he's done this, he'll sigh happily and say thank you, and move on to the next flock. This kind of birder is someone who rates gulls above and beyond all other birds. He's a 'larophile' – a lover of gulls (a more unkind name for him is 'gullard'). His idea of fun is a day sitting on a rubbish tip, sorting through 10,000 adult and subadult lesser black-backed gulls of the subspecies *graellsii*, looking for any wearing big, numbered (Darvic) leg rings so he can trace their life history later, and hoping all the time to find one among them of the subspecies *fuscus*.

If that's you, then you're a larophile. A guller. And you're probably off out gulling soon because it's your favourite thing to do in the whole world. You are happy to be out there all day at the reservoir or the harbour or the riverside or the rubbish-tip – just you, perhaps a like-minded friend or two, and lots of gulls.

Larophilia, found commonly in north-western Europe and especially Britain, is a curious condition. Not many other bird families seem to draw in their own special fanatical following – and it's particularly odd, given that gulls aren't really that inherently appealing – not cute, not colourful, certainly not sweet-voiced or particularly sweet-natured. But they do present a particularly tricky set of identification challenges, they do gather in big groups so there's lots to get your teeth into, and they are also relatively easy to watch and study – they offer a very real way to do 'scientific birding' for the amateur. Thanks to the larophiles, we know an awful lot about gulls these days, and it's also thanks to the dedicated larophiles of Britain that the British Bird List (as compiled by the British Ornithologists' Union) has 24 species of gulls on it and another one pending, as of 2015, despite there being only eight gull species regularly breeding in the British Isles.

Understanding larophilia means understanding birdwatching. I think that one of the misconceptions that the general public holds about birdwatching is that birdwatchers watch birds. Just sit there, watching them. Well, we *do* sometimes do that, if the birds are doing something interesting. Watching gannets plunge-diving into the sea is exciting, of course. Watching a mother mallard leading her ducklings out into the water is delightful. Watching a couple of white-tailed eagles locking talons in territorial battle is thrilling. Watching a coot awkwardly skating across a frozen lake is amusing. But birds spend a lot of time doing very little. It would be a very committed birdwatcher indeed who'd spend more than a minute watching a sleeping spoonbill, for example. Even if you've never seen a spoonbill before, the novelty of watching one standing on one leg, motionless, like a big white feather duster with its famous spoon-shaped bill tucked out of sight, would surely wear off pretty quickly. Every birdwatcher hopes to find something new to them, and/or something scarce or rare, of course, but that's in the lap of the gods. Once you've been birding for a while, most times that you go out you're not going to see any species that you've not seen many times before.

So, yes, when you're birdwatching, just watching birds isn't necessarily enough, and finding rarities doesn't happen that often (that'll be why they're rarities). This is where birders and twitchers diverge. The twitcher's interest lies in seeing as many species as possible, and to do this effectively you don't just go out looking for birds on your own – you go and look at the rare birds that other people have found. You go twitching, you twitch the bird. The slightly weird thing is that when you go on a twitch, often you're not actually looking for a bird. You're looking for a crowd of people with telescopes. As long as you can find them, you'll be OK because *they* will show you the bird (with luck). Twitchers rack up lots of miles on the road and in the air, and inevitably sometimes they don't manage to see their target bird – it's never, ever guaranteed. So they rack up lots of nervous tension too (hence the twitching).

Most birders have a bit of the twitcher in them, but it takes a certain type of person, and a certain level of wealth, to be a full-on, full-time twitcher. We birders who lack the motivation or means (or both) to do it have other ways of expanding our hobby. We like to identify each different species we see, and, if we can, work out the age and sex of each individual bird. (Not all birds have different plumages at different ages, and in many cases the sexes look nearly identical, but a practised eye can pick up subtle differences.) Such basic observations as these can get the mind working, and lead to deeper insights about the species as a whole. For example, the Swedish naturalist Linnaeus, who came up with the idea of binomial (two-part) scientific names for living things, gave the chaffinch the species name *coelebs* – 'bachelor' – after noticing that in winter nearly all the chaffinches he saw around were males. Much later on we learned that, in the far north of their range, many of the less tough chaffinches are forced to migrate south because there's not enough food for all of them in winter in those areas, and competitive squabbling is rife. Female chaffinches are, on average, a little smaller than males (20.9g, compared to 22.6g for the average male) so it's mostly females that get booted out of their breeding grounds, leaving one big bachelor party.

So, we note species and, if possible, age and sex. We might count how many individual birds we see of each sex and in each age class. We might note things like signs of breeding behaviour (singing, courtship, nest building, food carrying). We will perhaps make a list of what we've seen on an outing, and we might also have other, bigger lists. Most British birders keep track of all the species they've seen in Britain. Fiercely competitive twitchers discreetly size one another up by comparing the sizes of their British lists – you need to be packing 400 species at least on your BL to escape their scorn. Other list types include a world list (everything you've seen everywhere in the world, ever), year list (everything you've seen this year, either nationally or globally), patch list (everything you've seen on your favourite local birding spot) … the list of lists is limited only by your imagination.

Then there is the 'big day' list. This is when you set aside a day especially for seeing as many different species of birds as you can. This type of birding has evolved into something of a team sport – 'bird-racing'. In most kinds of bird race, you assemble a team of four and spend a day (up to the full 24 hours if you are really hardcore) searching a predesignated area (often a county) for as many species as possible. You can compete against other teams or against yourself, year on year. Some official bird races pit dozens of teams against one another and raise lots of cash for bird conservation groups.

My bird race experience is limited, but I used to do a 'big day' each January, as close to New Year's Day as I and my friends could manage it. We began at Nigel's house, in the conservatory overlooking his wooded East Sussex garden, breakfasting and birding at first light. With a bunch of little garden birds in the bag (hopefully including the local speciality – marsh tit), we'd head out to the coast, to sea watch from the cliffs at Fairlight and then scan the shallow coastal lagoons at Pett Level. By mid-morning we'd be at Dungeness. This was the best bit – Dungeness in January held so much potential for the unusual. Bitterns, smews, rare grebes and even divers … and also gulls.

Walking along a high shingle ridge towards the sea-watching hide on Dungeness beach, in the company of my three bird race pals, I was watching seaward rather than landward. The waves were high, foamy-crested. We'd already ticked off gannets and guillemots far out at sea, and now we were heading towards the 'patch' – the warm-water outflow that attracts a constant storm of gulls, and perhaps the odd skua if you're really lucky. Gulls were travelling parallel to the shore beside us, making their way to or from the patch. Herring gulls, great black-backs, common and black-headed gulls. As we got closer, one of my friends stopped to scan the swarm ahead and called 'Kittiwake!' It was important that we all saw the kittiwake – even on an unofficial bird race all team members are supposed to see the bird. So we all stopped and the finder of the kittiwake directed our gaze to where it was. I raised my binoculars and checked every gull that flew through their field of view, checking each one for those unmistakable kittiwake traits – a biggish dark marking behind the eye and another in front that gave it a frown, the grey neck shawl, the dipped-in-ink wingtips, the little notch at the tail-tip. The other two got it straight away but I could not, try as I might. Every gull I checked was a non-kittiwake. In the end I gave up and walked on … only to almost trip over a gull which was standing quietly and inconspicuously on the shingle ridge right in our path, clearly not feeling its best and taking a break from the mayhem. It was a kittiwake (well, of course it was), small and chubby, stumpy-legged, complete with shawl and inky wingtips and beautiful frowning face, and it went on our list. I only hope it also went back to sea soon afterwards.

Those January bird race days also yielded me my first glaucous gull and first Iceland gull, both of them at Dungeness. The former was hanging around the fishing boats on the beach, while the latter was loafing on a shingle island on one of the lagoons. Both of these gulls were first-winters, or, to use the more modern terminology, 2cy birds, born a few months before. They were very alike in colour scheme – creamy-white with lots of pale brown speckling, so they looked ghostly alongside the young herring gulls and great black-backs, but in shape and build they were very different. The glaucous was a great big robust thing, the Iceland slim, slight and elegant. We are very used to finding out that similar-looking birds are actually not closely related at all, but in fact DNA comparisons show that these two lookalikes are close cousins as well. I treasured these sightings as my 'firsts' of their kind, but also because they boosted our bird race score for the day by one apiece. That's the curious nature of birdwatching and birders – seeing a new bird for the first time is such a joy, but it's also just a number on a list.

In spring 2015, I went to visit some birding pals, Mike and Hazel, up in the north-west of England, and as the date of the trip neared, I found myself getting twitchy. For nearly the whole winter, there had been a young laughing gull hanging out at the marine lake in New Brighton, on the Wirral, close to where my friends lived. They had been to see this bird several times, photographing it as its plumage gradually changed through the winter. Looking at these pictures, I fell in long-distance love with the laughing gull – in its smoky colours, with its funny gangly body, this American wanderer was so utterly different to all of the British gulls I knew so well. I hoped ardently that it would stay put long enough for me to see it in the feather.

When we arrived at the lake in the morning, there was no sign of the gull. There were some gulls over on the Wirral's idea of a beach, wandering lethargically about on the exposed muddy shoreline – some black-headed, some herring gulls … but no laughing gull. It was supposed to hang out on the pontoon here, along with a flock of turnstones. I'd seen the photos to prove it, and the bird had been reported the day before – but now, no gull, nor turnstones. The pontoon was spattered with bird poo but devoid of birds, save a lone pied wagtail strutting pertly along, snapping up the flies attracted to the aforementioned bird poo. I fought to transform my disappointment into patient optimism. My friends suggested going for a coffee and trying again afterwards. I said that they should do exactly that, but that I would stay put and watch the lake. Much as I fancied a coffee, the idea that the bird might come and go while I was working my way through a large flat white was more than I could handle. Sacrifices have to be made when one is on a twitch. So off they went, and I took up position on the side of the big, concrete-framed lake to wait.

It was pleasant here, with a slight wintery bite to the early April air. As a big fan of seasides of all kinds, I was interested in the nature of the coastline here, but found it a little disappointing – the vastness of the mudflats meant that the washing waves were slow, sluggish, seeming to seep in and out without any photogenic drama. The tide was high but there was a little exposed mud, which offered good feeding ground for waders, and there were indeed a few redshanks here, pecking around in search of saltwater-loving worms and the like. I turned my attention from the shore to the lake, and watched a hefty 2cy/first-winter herring gull wheel overhead. It was a sunny morning and the light caught the young gull's tawny tones quite beautifully. I lowered my binoculars and noticed that another gull was approaching from the north – but it didn't look tawny, and it didn't look hefty. I raised my binoculars on the target and I was looking at the laughing gull. My heart rate shot up as the bird sailed closer and lower. It was all charcoal and dusky, all long wings, elegant neck and dolefully downturned bill. I switched from binoculars to camera and began to photograph the bird as it homed in on the lake, making a low circle before landing neatly on the pontoon and settling its plumage.

I sent my friends a quick text – 'Laughing boy/girl is here' – and found the closest part of the lakeside to where the gull was standing. It was a morose-looking bird, belying its name in its hangdog demeanour, formed by that droopy bill, an equally droopy gape line, and the smudgy face markings that conjured up an expression of sorrowful resignation. Its hunched posture only added to that impression, but then my friends joined me and the gull perked up. They had hurried over via the supermarket next to the coffee shop, where they had bought some chunks of the cheapest fresh fish they could find. Hazel hurled a lump of fish down towards the pontoon where the gull sat. With typical larine gusto it dashed over, grabbed the fish and immediately launched itself skywards, circling the lake while gulping down the awkward mouthful. In flight, at close range, I could see the sparse traces of brown juvenile plumage in the inner parts of its wings. Hazel chucked another bit of fish onto the pontoon. The gull dropped down to collect it, then flew up, fish in bill, and away towards the sea. We waited a little while but it didn't return. What would become of this young bird? Whether it would ever find its way home to North America we could never know, but at least it had found generous hospitality here in England.

Going to see the laughing gull qualified as a twitch. I knew the bird was (probably) going to be there. No other twitchers were at the site that morning, but that was only because the gull had been at the lake for months, and the keenest twitchers had already been, back in early February when the bird first turned up. What birders long for most of all, though, is to find a rarity of their own – that heart-stopping moment when you realise that, completely out of the blue, you've found something really unusual.

For a larophile, spending lots of time at gull-filled places, there is a real chance of 'self-finding' a scarce, rare or even 'mega' (that's really, really rare) gull, which the twitchers may then travel to see. The slaty-backed gull (a mega if ever there was one) that was found at the landfill site adjacent to RSPB Rainham Marshes, London, in January 2011 appeared late on a Thursday afternoon – not long before sunset. It was still around on and off on the Friday, delighting a dozen or so twitchers who'd travelled over on hearing the news. However, many more couldn't get to the site until the Saturday, pesky things like jobs and childcare getting in the way. So it came to pass that about 1,200 very twitchy birders showed up on the Saturday morning. These unfortunate folk endured a rather harrowing day in ankle-deep mud, pressed up against the landfill site's fence cheek to jowl with other twitchers (probably the only thing saving them from mild hypothermia), all searching for the slaty-backed gull, and all to no avail. One of my birding friends who went on the Saturday described it as the most miserable day he'd ever spent in the field – with 'field' very much not being the operative word. When the average person thinks about birdwatching, they probably don't imagine anything like this. Happily for the twitching world, the gull did make a few more appearances through February, and most of those eager to see it did manage to do so in the end, including my disgruntled friend.

Going to see the slaty-backed gull just didn't appeal to me. I realise that with that statement I confirm my status as a non-larophile, as well as a non-twitcher. Had it turned up somewhere a bit more pleasant-smelling, or a bit more easily reached by public transport from where I live, or had there been a chance of good, close views (and photos) then I might have done – but the gull was never seen at really close quarters by anyone. So I don't really envy the people who did go to see it – but I do rather envy the original finder, Dominic Mitchell.

That thrill of unearthing a really rare bird – in this case a first for Britain – that's something that every birder would understand, though it can be a double-edged sword.

Finding a rare bird is, by definition, difficult. Identifying it is sometimes very easy. It's sometimes not. Some birds are highly distinctive, even if seen poorly and briefly. Finding the rarity is part of the picture but to really win the admiration of your fellow birders you have to nail the identification as well. I needed a hand with the ID of the rarest bird I've ever found – a juvenile Baillon's crake (think small moorhen-like thing, streaky-brown and furtive), once again at RSPB Rainham Marshes. It was September, I'd been photographing hobbies and was feeling disgruntled about not having managed to see what I'd gone to the reserve for – the rare southern migrant hawker dragonfly. I took myself off to the tower hide at the far end of the reserve and gazed out across the sweep of marshland and meadow, before something caught my attention closer at hand. A little bird had crept out of a sedge clump (and, in short order, crept back in again) directly in front of me, as I sat alone in the hide. I could see that it was of the crake/rail ilk, but it was some way off and in poor light. Thankfully, I managed to fire off a couple of shoddy photos, and when I returned to the visitor centre I showed the pictures to someone more knowledgeable than me (the marvellous site manager Howard Vaughan). We put the pics up on his big screen to confirm which kind of crake it was. Although I felt deep shame for not nailing the identification on my own, at least I'd known it was a crake, special enough that I knew I should take some photos, however appalling. Without those pics, I'd have 'lost' that bird, and would not have had the glory of writing up my sighting for the *London Bird Report*, but I still feel an awkward mixture of pride and embarrassment when I think about that Baillon's crake.

When Dominic Mitchell found his slaty-backed gull, he too was able to obtain some photos but the bird was far away and among a big crowd of other gulls, so they were not the crystal clear images you'd want under such circumstances. As a very experienced world birder and a true larophile, he knew what he'd seen, but the bird's identification was still fiercely debated among the many, many twitchers who'd seen the photos but not the bird itself. Reading some of the comments posted about it online at the time, you definitely sense the presence of sour grapes here and there. Let us hope that those unhappy twitchers casting aspersions over the identification were among the many who did eventually see the slaty-backed gull when it deigned to reappear some days later.

As it happens, one of my best ever 'self-found' birds was a gull, and I did identify it at a glance all on my own. *And* I found it while looking at lots of gulls in a fishing harbour, so I suppose that was a genuine larophile moment for me, even if I'm not a real larophile. This happened in March 2014, when I was visiting my favourite Cornish village, Mousehole. There had been some seriously stormy weather that winter, enough to breach harbours, smash the windows of seafront homes with wave-flung rocks, and to knock a Devon train line into the sea. I'd walked from Mousehole to Penzance around the sweep of Mount's Bay, and seen the destruction for myself – the shore and promenade in Penzance were still liberally strewn with debris. The after-effects on the birds were also extraordinary. I'd stopped off at Newlyn Harbour on the way and counted seven great northern divers fishing in its calm waters, along with a number of shags, guillemots and razorbills. All of these birds would normally be well offshore in the relative calm of the bay – it takes very bad weather to drive so many tough seafarers into an actual harbour. There were also lots of kittiwakes among the usual gulls, and further along, at the far end of Penzance, I saw a young glaucous gull riding the still-restless sea – huge and magnificent in its creamy plumage.

On the return walk, I decided to look in at Newlyn once more and this time checked out the short, square segment of water furthest from the open sea. Here, around the massive storage sheds close to the water's edge, was a flock of turnstones, fat and dusky little waders on hectic orange legs. They all hurried over to mill around my feet for crumbs when I opened my bag of crisps. I sat down and watched them for a little while, enjoying their skittering activity and tuneful mutterings, then looked beyond them to the harbour itself. Several gulls were bobbing on the silky-looking water, and one of them instantly caught my eye. It was the smallest bird present in the group, which was otherwise composed of herring gulls and a few great black-backs. It was long-winged and dainty-headed. In coloration it looked rather like a mini adult herring gull but its grey bits were even more frosty-silvery, its eyes were dusky rather than glacial pale, and its wingtips, rather than black with white spots, were light grey and white in an indistinct pattern. It was close enough that I could see its legs through the water and they were pink. I knew that I was looking at something a bit special – specifically, an adult Kumlien's gull – *Larus glaucoides kumlieni*. I started to take photos of it, as it tilted its head my way and gave me a thoughtful once-over with that curious golden-grey eye.

This gull, nonchalantly sharing a harbour with the locals, had come a long way. Kumlien's gulls live in Canada. They breed in arctic eastern Canada, and spend the winter further south, along the eastern coast of the USA down to Florida. So at some point, this bird had crossed the full width of the Atlantic to reach south-western Cornwall.

That knowledge made it a very exciting find for me. However, there are a few Kumlien's gulls reported every winter in Britain – they are classed as 'scarce' rather than 'rare', which means that records do not need to be assessed by the British Birds Rarities Committee (the official adjudicator of rare bird records in Britain). There are also a number of problems with identifying Kumlien's gulls – they are highly variable, and particularly tricky to ID in subadult plumages. The final issue with this bird is that no one is quite sure what it is…

Most people who care about such things class it as a subspecies of Iceland gull. However, in both its looks and its distribution, Kumlien's gull sits neatly in between our 'usual' subspecies of Iceland gull (which breeds in Greenland, and is very pale with pure white wingtips), and Thayer's gull (which breeds in northern Canada and is darker, with black, white-spotted wingtips). So perhaps Kumlien's gull isn't a subspecies at all, but a 'hybrid swarm' between Iceland gull and Thayer's gull. That would account for its notorious variability – some

Kumlien's are much darker than others. But there is also a third way – some ornithologists support a full species status for Kumlien's.

In a final and probably decisive (well, we can hope) taxonomic twist to this story, the American Ornithologists' Union decided in 2017 that there was no clear distinction between Thayer's, Kumlien's and Iceland gull and that all three should be 'lumped' together as a single species. It would retain the English name 'Iceland gull', but within this new and improved Iceland gull there would be three subspecies – *Larus glaucoides glaucoides* (the original Iceland gull), *Larus glaucoides thayeri* (Thayer's gull, demoted from species to subspecies) and *Larus glaucoides kumlieni* (Kumlien's gull, finally defined as a subspecies rather than a hybrid form or a full species). The British Ornithologists' Union is likely to follow suit with this classification very soon. This makes things simpler, for sure.

'My' Kumlien's gull showed no outward signs of wrestling with an identity crisis, and indeed in terms of its plumage it was a classic example of *kumlieni*. I spent some time watching it, but it wasn't doing a lot, and even when you are eye to eye with a self-found rare (oh, alright, scarce) bird, the thrill dies down quite quickly when all said bird is doing is floating around and staring into space. So I went off on my way. I considered reporting the sighting to the county birding groups, but found when I got back to the cottage in Mousehole and checked online that someone else had already done so – another birder or birders must have 'found' 'my' gull too, either before or after me.

I have two more rare (OK, scarce) gull tales to tell. The first concerns probably the laziest twitch of my whole birding life – I waited almost 11 years to go and see this bird. As I mentioned earlier, the ring-billed gull is the most frequent North American gull species to visit Britain, and the records include several regularly returning birds. One such bird made its first appearance (or at least was first noticed by a birder) in the year 2000. The finder, Steve Arlow, discovered the adult male ring-billed gull in March, in Westcliff-on-Sea in south Essex. The gull had departed by April, but next autumn he was back, and stayed through the whole winter. Thereafter, he visited each winter and

145

many local birders went to see him at least once a year. He acquired a nickname, 'Rossi', after the ice cream parlour that marked the seafront spot where he tended to hang out. His appetite for bread meant that it was easy to entice him close, along with his entourage of black-headed gulls.

I went to see Rossi in January 2011. Why I left it so long to visit this relatively local celebrity, I'm not sure, but he didn't hold it against me – it took only a few minutes of searching to locate him. My first sighting was when someone more savvy than me emptied a bag of bread crusts onto the promenade, some way further along from where I was standing. Gulls and pigeons immediately came flying in from all directions and I had a good, though distant, view of them all. The gulls were nearly all black-headed, but among them was a particularly big and silvery-looking individual. I raised my camera just as this burly bird barged past the rest to grab the first and biggest crust, alighting on the stone wall with wings raised and spread as if to fend off his rivals. Through my big lens I could see the neat black bill marking, the severe expression conveyed by his pale eyes and the yellowy legs – a combination that set him apart from all the British gull species I knew.

I walked along the beach for a while, and among the black-headed gulls floating on the sea I found a couple of Mediterranean gulls – a nice bonus addition to the day list. Then I found Rossi again, resting on a breakwater with a number of the ubiquitous black-headed gulls. He stood among them like a king surrounded by minions, and this time he was close enough that I could see the blood-red ring of skin encircling his pale lemon-yellow eye. He was a fine figure of a gull indeed, not quite as big as a herring gull but with the same sturdy proportions and don't-mess-with-me attitude. I took one step too close and the black-headed gulls took flight, but Rossi stayed put, staring me down.

Where Rossi went in the summer was and is a mystery. Steve Arlow noted in his blog that Rossi usually appeared at the end of August and was usually gone before April. Presumably he (Rossi, not Steve) felt the usual seasonal stirrings and went forth in search of love. It seems unlikely that he returned to North America – more likely he would head somewhere more northerly in Europe. Maybe he actually had a mate somewhere – another ring-billed gull, or (more likely) a common gull who was up for a bit of hybridisation. Whatever occupied him from April to mid-August, he would be back at Westcliff without fail … until he wasn't. It turned out that 2011 was the last year he was seen, so I got there just in time. He was at least 14 years old when he disappeared – a lifespan not to be sniffed at, especially for a bird so drastically displaced from his natal home.

In March 2015, I twitched another ring-billed gull. This one, though, was a mere baby, in its first winter and its second calendar year. It had been present at a lake in Falmouth, Cornwall, for about three weeks already. The lake in question was a big one, shaded in places with trees, circled by a pleasant footpath, bestrewn with ducks and other wildfowl, and popular with families. The people came to feed the ducks, and accordingly the young ring-billed gull had worked out that human presence often meant free food. This made for a remarkably easy twitch – the moment I reached the shoreline I saw the gull in the air, flying towards me in a hurry. There followed half an hour of sheer joy as I watched and photographed this exuberant and (to me, at least) beautiful and distinctive gull as it flew, swam, screamed, postured and generally disported itself around the place.

This bird looked, structurally, rather slighter and daintier than I remembered Rossi to be. He was more herring gull-like, while this bird put me more in mind of a common gull. So I decided it was a female. Plumage-wise, she was a bit chaotic, with heavy brown speckling over the bits of her plumage that were destined to be white in adulthood, and some light frosty-grey feathers mixed in with the tawny-brown dappled ones on her back and in her wings. Her wingtips and outer wing edges were solid black, her tail dark-banded. She had not yet begun to moult any flight feathers, but my photos show that the central two tail feathers were much whiter than the rest and their dark band narrower and neater – they were newly grown second-winter feathers, while the rest of her tail was still that of a juvenile.

It was her face that I found most arresting – a cute round head with big, very dark eyes accentuated by a dark frowning smudge at their inner corners, and a rather hefty pink bill with a thick black ring near its tip. When some people arrived to chuck out some bread, she set about grabbing it with gusto, and when there was no more bread to scrounge she swam about energetically, pausing often to throw back her head and yell loudly at no one in particular, or take off and fly restless circuits of the lake.

I've seen a few other out-of-the-ordinary gulls over my years of birding and occasionally twitching, but these are the ones I remember with most affection. As a lifelong fan of gulls, I'm probably more likely to want to twitch a gull than most other birds, and as a birder, I'm very happy to see and (try to) identify gulls everywhere I go. But I draw the line at going to the rubbish dump for an all-day session, and I draw the line at spending more than half an hour staring at the same flock of gulls. That's why I'm not a real larophile.

If you want a true insight into the nature of larophilia (and also to learn an incredible amount about the finer details of identifying northern hemisphere gulls), I recommend you head over to the Gull Research Organisation's website, at www.gull-research.org. This extraordinary resource has month-by-month descriptions of all the gull species from fledging age to adulthood, with numerous annotated photos to explain exactly what the subtle key pointers are in each case. The home page has little mugshots of its 34 authors, an array of men (yes, they're all men) of varied nationalities across the northern hemisphere – most are northern Europeans but there are others from Israel, Canada, Korea and southern Europe.

To be a real larophile, you have to talk the talk and walk the walk. Like most special interests, gulling has its own special vocabulary. The terms include affectionate nicknames for certain gull species (e.g. 'Smith' for *Larus smithsonianus*, and 'glonk' for the glaucous gull) but a lot of more arcane stuff too. Let's start with some of the basics – like feather names.

OK, not every individual feather that grows on a gull's body has its own special name. That would be a level of nerdiness that even the gulliest of gullards could not countenance (I hope not, anyway). But some of the groups or tracts of feathers on a bird's body contain only a few large feathers. Take the primary feathers. These are the outermost flight feathers, the longest feathers of all in the case of gulls and most other birds. When the bird is flying, the primary feathers make up the trailing edge of the 'hand' of the wing (as opposed to the inner or 'arm' part of the wing). When the wings are closed, the primary feathers are folded together to make a long narrow point at the rear of the body – this is often mistakenly called the 'tail', and usually projects some way beyond the actual tail. Gulls have 10 primary feathers and they are known as p1, p2, p3 and so on, with p1 being the innermost (closest to the body) and p10 the outermost. The patterns on the tip of the outermost five or so primaries are often very important in identification.

The parts of the feather also have special names. The hard supporting bit that roots into the skin and goes up the feather centre is the 'shaft', and the flexible parts of the feather either side of the shaft, formed by softish vanes that hook together, Velcro-style, are the 'webs'. In the case of the flight feathers, the web that would sit closest to the body when the feather is attached to the bird is the 'inner web', and its counterpart on the other side is the 'outer web'.

The markings on an outer primary feather have their own special names too. In the case of most large gulls, the inner part of the primary feather is grey, then there's a narrow white edge separating the grey from the outer part of the feather, which is black. Then there are some white markings within the black bit. If there is white at the very tip of the feather, that's just called a 'tip', but a white spot within the black is a 'mirror'. Where the white-edged grey of the inner part of the feather 'eats into' the black of the outer part, that's called a 'tongue'. Put all that jargon together, and you get horrifying sentences like 'P10 with white tip and small white mirror, extensive tongue on outer web.' However, if you read gibberish like this while looking at photos of gulls with their wings spread, it does get a little less bewildering (I promise). If you want to exercise your incipient inner larophile, pick up a moulted gull primary feather when you find one, and try to work out which species of gull it belonged to and where in the p1–10 sequence it belonged.

The secondaries are the flight feathers that follow on from the primaries, occupying the trailing edge of the inner, 'arm' part of the wing. They are counted the other way round to primaries, with s1 being the outermost one. The total number of secondary feathers varies from species to species (from 16 to 23). The secondaries are not usually visible (or only barely visible) when the gull isn't flying, and they tend to be uniformly marked, so they're of much less use in identification.

The innermost flight feathers are called the tertials. In the spread wing they just continue on from the secondaries but they form a distinctive grouping when the wing is closed. You can usually see up to six of them in the folded wing, and they are counted downwards in this view, so t1 is the one at the top. The pattern on the tertials can be helpful for identifying young gulls in particular. For example, on a first-winter herring gull, the tertials are pale brown with a central dark brown marking shaped like an oak leaf (i.e. with a wobbly edge), while in a first-winter lesser black-back they're almost completely dark brown with just a narrow pale-brown edge. In adult gulls the tertials are the same shade of grey as the rest of the wing but with white at the tip, which means that when the wing is folded those tips form a white band in between the grey of the inner part of the wing and the mostly black or dark-grey wingtip, formed by the coloured primary tips (except in the 'white-winged' species, which have pure white primary tips). The broadness of this white band can be important in identification. So too can be the angle between the tertials and the primaries on the folded wing. This angle is called the 'tertial step' because in pronounced cases there is a definite step shape formed where tertials meet primaries. The size of the tertial step can also be helpful in identification – for example, the glaucous gull has a big one, while the very similar (though smaller) Iceland gull doesn't really have one at all.

Collectively, the flight feathers in a gull's wing (primaries, secondaries and tertials) are called the remiges. The long tail feathers (which form the main 'fan' of the tail, as opposed to the shorter ones at the base) are called the rectrices and they are also, debatably, flight feathers – their function is to steer and brake when the bird is in the air. Gulls have twelve rectrices, with the central pair lying on top (when you look down onto the tail) and the others stacked below in pairs that sit together in parallel when the tail is not fanned out. Adult gulls of most species have pure white tails, but juveniles usually have a dark band almost at the tail-tip, and sometimes narrower bands behind. The pattern is often helpful for identification – for example, the tail band of a juvenile yellow-legged gull is blacker and more distinct than that of a herring gull. The tail band and other tail markings diminish with each moult – a few smudges of dark coloration in the tail-tip are often the last traces of youth in an otherwise adult-plumaged gull.

Most of the details of gull identification lie in flight feather and tail patterns but there is more. Patterns on the shorter, inner wing feathers – the coverts – can be useful in young gulls. The same is true of the axillaries – the long slender feathers in the 'armpit' which are revealed when the bird raises its wings. Head markings are often important, but body plumage is usually less so. The dark head developed by some (mostly) smaller gulls in summer is called the hood, and its shape varies between species. Most hooded gulls also have white markings above and below the eye – these are the eye-crescents and they vary between species in how prominent they are (for example, they are big and obvious in the adult Mediterranean gull but much less so in the black-headed gull).

Then there are the 'bare parts'. These are the unfeathered bits of the gull – its bill, eyes, and legs and feet. The bill's colour and pattern varies with age and also by season – most adult gulls have duller-toned bills outside the breeding season, though the black-headed gull is an exception, having a bright red 'winter bill' that darkens to almost black in summer. Gulls that have pale eyes in adulthood have very dark eyes in their first year or two of life. Leg colour also tends to brighten with age. Bill shape and leg length are also useful identification pointers, as is the apparent size of the eye (the Caspian gull is often rather uncharitably described as having 'pinprick' or even 'piggy' little eyes, compared to other large gulls). Even the bill itself has a special little lexicon to get to know. The top edge is the culmen – its curviness varies between species – and the pointed projection on the bottom edge, near the tip, is the gonys – its shape, or gonydeal angle, is important in identification. The red dot on this part of the bill in larger gulls is the gonydeal spot.

Some of the other terms you might find are 'eye-spot' (the dot behind the eye on the cheeks of winter-plumaged 'hooded' gulls), and 'shawl' for the dusky wash across the neck on, for example, kittiwakes. Feathers might be marked with 'notches' (where a pale outer edge eats into a darker centre) or be 'oak-leafed' (showing a curvy pattern like the shape of an oak leaf). Some gulls show a rather puffy undertail area when standing, which might be described as a 'ventral bulge', or more unkindly as a 'saggy belly'. A pale patch on the inner primary feathers is a 'window'. The coloured circle of skin around the eye is an 'orbital ring'. Finally, certain colour descriptions come up a lot when larophiles talk about gulls, such as 'bubblegum-pink' for the bright legs of an adult Thayer's gull or the bill-base of a young glaucous gull.

Size and shape are also important with identification. Because gulls are usually seen in big mixed groups, you will often be able to directly compare the sizes of different species standing side by side, though beware that there's a lot of size variation within one species. A small female great black-backed gull can weigh as little as 1.29kg, which is only a shade heavier than the biggest male herring gulls. In general, shape will always be more useful than size as a guide to identification, and the same goes for pattern over colour, because our impressions of size and of colour are much more likely to be distorted by outside factors like light and perspective.

It's not all about looks, either. Larophiles will also be paying attention to the gulls' behaviour (unless they are all asleep) – their calls, how they interact with other gull species, their body postures and so on. The large white-headed gulls all show differences in their 'long-call' (that's the one that they begin in a bowed-over posture and then throw back their heads to deliver a series of loud gargly yelps) – not only in sound but in movement. Caspian gulls, especially big males, are often very noisy and strikingly aggressive towards other gulls (perhaps it's all down to insecurity over the size of their eyes).

There are all of these things to weigh up when trying to identify a gull, and because there is so much individual variation, you will rarely be able to make a confident call on a more ambiguous gull without checking out several of them. Appearance also changes through the seasons – not just during moult, but gradually over the months before moulting even begins. Dark colours bleach paler, and feather tips wear away, making those diagnostic wingtip patterns suddenly less reliable. Then there is the matter of hybrid birds, which inherit a completely unpredictable melange of traits from their two parent species, and one last spanner is hurled into the works when you add in birds with pigment abnormalities, such as leucism which turns normally coloured feathers white.

This level of potential confusion really is enough to make you throw your hands aloft in despair. I've met some birders who flatly state that they 'don't do gulls', and many more who say sheepishly that they're 'not very good at gulls'. Gulls, along with warblers and waders, are definitely one of those tricky groups that newcomers to birdwatching find the most difficult. Case in point – let's take a quick look at the results of the RSPB's Big Garden Birdwatch survey from 2019. This survey, now in its fourth decade, is for everyone with a garden. No level of birdwatching expertise is expected – the survey pack includes identification charts for the birds you're most likely to see over the survey period (one hour on one particular weekend in January). Its simplicity and child-friendliness makes it the most popular bird survey in Britain, with nearly half a million households submitting results in 2019. However, when it comes to gulls I strongly suspect the data is in error. Three gull species are in the top 40 most frequently seen species – the herring gull at number 32, black-headed gull at number 30, and common gull at 27. But I'd stake my binoculars on the bet that many of those common gulls are actually herring gulls. Common gulls would be, in my experience, much less likely to come and take food from gardens than either black-headed or herring gulls. The trouble is that lots of non-birders assume that the gulls they see most often must be common gulls, simply because of the name. However, getting tough about correct gull identification would surely alienate some of the Big Garden Birdwatch participants, and put them off getting involved at all.

Gull identification is a minefield, for sure, but the existence of larophiles shows that it's a minefield that's wildly inspiring and exciting to some. So who are these people whose blood is roused by such things? What could turn a respectable ordinary birder like me into someone who would hang out at rubbish dumps and fill notebooks with fevered script about tertial steps and gonydeal angles, inner and outer webs, long-calls and threat postures, and tips, tongues and mirrors? What marks out someone who might have the larid ID chops to find a new gull for Britain someday?

I talked to some of the committed 'gullers' that I know about what got them so hooked. They answered, to a man (yes, they were all men) that the challenge of identification is the biggest draw, though several also mentioned that they enjoy finding and reporting Darvic-ringed birds, and that they like the fact that gulls often congregate in places that are often free or mostly free of people. Even though it's generally acknowledged among birders that gulls are 'difficult', none of my larophiles really cared that 'being good at gulls' might impress others. And when I asked what their non-birding friends make of it all, the replies included such words as 'confusion', 'scorn', 'strange', 'odd' and 'weird'.

Here's some of these gentlemen in their own words. Harry says he enjoys gull watching because of 'the chance of finding something scarce or rare in unpromising locations', and that 'gulls offered my best chance of 'quality' birds in the local area.' I hadn't actually thought about this, but it is much more difficult to find rare or unusual birds if you don't live in the wilderness or close to the coast, but there probably is a gull roost somewhere not too far away from where you live and it's as likely to attract a rare gull as any other gull roost.

Peter, whose fortnightly gulling trips at home are augmented by gull-searching overseas, spoke about the joy of challenging oneself to gull identification even on a quiet day: 'One can have as much fun finally clinching a juvenile lesser black-back amongst a flock of herring gulls as finding a Caspian or yellow legged gull.' He also has a fondness for the birds' varied but always engaging characters: 'There is rarely a dull moment when gulling ... the interesting diversity of plumages and the intelligence and cheekiness of the commoner species makes gulls one of my favourite bird families.'

David, whose work includes leading tours and birding walks, and generally engaging the public with birds and birdwatching, wins over the sceptics by explaining that 'these birds in the winter have possibly travelled thousands of miles to be here.' He enjoys seeking out Darvic-ringed birds and finds that his clients are particularly intrigued by these, especially any individuals that hadn't been observed since they were first ringed, until now.

Yoav does most of his gulling in Israel and was the finder of Israel's second ever Franklin's gull, an American species that had travelled at least 11,000km. Like another couple of people who talked to me, his favourite species is the magnificent Pallas's gull or great black-headed gull – unlike the Brits, he gets to see this species on a regular basis. Another species that came up as favourite for many is the Mediterranean gull, an exceptionally good-looking species, and one that's pretty easy to identify in all its plumages. I took this as proof, if it were needed, that larophiles do prize sheer beauty as well as fiendishly difficult identification challenges.

Michael does a lot of his gulling down at Dungeness, bringing food for the gulls that loiter around the fishing boats. He has also sought gulls out abroad, including a day-trip twitch to Italy to see a grey-headed gull (a wanderer from Africa) at Bari, in January 2015. I thought this was perhaps a bit much, but then I visited his blog (check it out – birdingthedayaway.blogspot.com) and suffered crippling envy at the beautiful photos he took of the bird, a gorgeous adult with scarlet legs and bill and plumage every shade of a seastorm. Michael enjoys the peace and quiet, from human noises at least, that gulling can bring: 'Gulling and the locations where gulls roost and feed generally are free of other people, and that makes it much more enjoyable.' When I asked him what his non-birding friends think of his hobby, he replied, 'Oh, they think I'm a proper weirdo.'

Stuart lives in Wisconsin, USA, and says that one of the reasons he likes gull-watching is because there's 'not much else to watch in Wisconsin in winter! Plus, they stand still and can be scoped.' That first part is not strictly true – where he lives is famed for bald eagles in winter, but Stuart isn't as bothered about them. 'When I reply to the inevitable "watching the eagles?" with "no, the gulls," there is a stunned silence,' he says. He highlights the value of the online community and social media for sharing photos and learning more about identification, mentioning the North American gulls Facebook group in particular. 'I posted a query on a potential slaty-backed that was eventually shown to be a 'Great Lakes gull' (American herring gull × great black-backed gull hybrid) and it was a brilliant thread – opinions from the US, Europe and even Japan – and it was a great learning experience.' He adds that 'gull-watchers know they are weird, so the online world is a very welcoming one!'

Talking to people who love gulls even more than I do has got me thinking. Larophilia as it is today has really only been a thing since the turn of the millennium. I think one of the most key moments in its development was marked by the publication, in 2004, of *Gulls of Europe, Asia and North America*. This hefty tome, by renowned gull expert Klaus Malling Olsen and illustrated by the insanely talented Hans Larsson, was revolutionary in lots of ways and is still the go-to gull reference book for us Northern Hemisphere-ites. There is the detail and extent of its coverage, both in word and image, with a vast array of photos to go alongside its glorious painted plates (often four or more for each species). Then there's the fact that it draws upon DNA analysis for the first time to talk about the relationships between different species, and to argue for the 'splits' of certain forms into new full species. Armed with this beast of a book, many birders felt emboldened for the first time to really get into the deep detail of gull identification, rather than simply ignoring all gulls that weren't clear-cut adult-plumaged examples of their species.

Growing knowledge and understanding about gull identification revealed lots of new stuff about where gulls can be found at different times of the year, and new ways to work out the identification of the trickiest individuals. Now, there's social media to spread the word further and wider. I've signed up to quite a few hardcore gull-related Facebook groups, and they are among the most active of all the wildlife groups I frequent. 'Western Palearctic Gulls' has 4,677 members at the time of writing, and 'North American Gulls' has 8,415. 'UK Gulls' has 1,659 members, and for those new to the joy of gulling there is the pretty new but fast-growing 'UK Gulls 4 beginners' with 1,383 members.

A look on Twitter reveals lots of gull-related accounts, their creators highlighting their gull-related tweets with hashtags such as #gulls, #TeamGull and #laridaerule. The spectacular www.gull-research.org website has a chat forum, with 1,867 posts in its 'identification' section. The big and busy www.birdforum.net has a dedicated subforum just for gulls – no other single bird family is given this honour. Even on Instagram, which I consider the most frivolous of the widely used social media platforms, the hashtag #gulls brings up 122,674 posts, and #seagulls 938,025 (though to be fair, use of the 's' word means we are probably moving away from true larophilia).

New gull field guides have also come along, such as *Gulls Simplified: a comparative approach to identification*, by Pete Dunne, which tackles North American species and seeks to make the rather dense language of larophilia more accessible to the newbie. There is also now a slightly pared down, photos-only gull ID book by Klaus Malling Olsen – *Gulls of the World: a photographic guide*, which adds in the relatively small number of southern-hemisphere-only species that were missing from the original. You can buy these for e-readers too, of course, so you can carry all that info around in your jacket pocket.

So there's no doubt that gull identification has become easier and more accessible, and that is surely one of the main drivers behind the surge of interest in these birds. But there's another, less happy side to it as well. A couple of the gullers I talked to mentioned that they watch gulls over and above other birds because other birds are hard to find, and that is increasingly the case, almost everywhere in the world. Lots of gull species are also not doing very well, but, as we know, gulls as a group are better than most at coping with the havoc humans are wreaking on the natural world. That's why it's still relatively easy to find gull gatherings in places where other birds – indeed, other wildlife – are fast dwindling away. We can be gull fans, we can be larophiles, but we need to keep an eye on the wider world too, because no living things, not even gulls, can thrive for long without their support network – a healthy ecosystem.

CHAPTER 7

Moving On

Black-headed gull

THE EARLIEST FOSSILS of any kind of gull date back at least 23 million years. That's some 13 million years before our own lineage of hominids split from the ancestors of the chimpanzee, so it's not really fair of us to characterise gull-kind as a bunch of cheeky upstarts. They've done pretty well thus far. It couldn't last, though. Today, we and almost all other living things face one hell of a crisis, though unlike all the others, our species has at least a chance to do something about it. The last mass extinction on Earth occurred about 66 million years ago and wiped out some 76% of all species around at the time. In the one before that, 250 million years ago, more than 90% of all living things were put to the natural selection sword. A new mass extinction is underway right now, caused by human activity. While we don't know yet how bad it will be in the end, we're currently losing species at 100–1,000 times the 'normal' rate. Gulls aren't immune to this – and neither will we be, in the end. At least we know from prehistory that some traces of life *will* almost certainly survive such events, and life is very likely to go on without us in one form or another.

Contemplating the future is probably not an activity that takes up much time in the average gull's day. Trying not to die and trying to pass on your genes are the two overwhelming life priorities if you're a gull (or nearly any other living thing, really). Even the gulls that you'll see standing about on various beaches in winter, apparently doing nothing whatsoever for hours on end, are actually busy furthering those ambitions. If you're not immediately hungry and it's not the breeding season, then doing as little as possible for now will preserve your energy and help your body and plumage stay in good condition, leaving you in a better position to forage later on, or escape a predator, or seduce another gull, or whatever you need to do to further those two goals of 1) survive and 2) procreate.

Certain physical traits and behaviours are good for survival and good for breeding, and if those things happen to be genetically determined (rather than learned or acquired during life), then they will be passed on to any offspring a gull produces. That's how natural selection works, and how evolution works. Every animal's body and quite a lot of its behaviour is determined by its DNA. This is the molecule that is in charge of making every different kind of protein – the fundamental building blocks of the body. DNA lives in the nucleus of each cell, and is organised into long strands called chromosomes. Each bit of a chromosome that makes a particular protein is called a gene. Animals inherit their genes from their parents, but some genes also change (mutate) during the cell division process prior to the formation of egg cells and sperm cells.

So each animal's genes are a mixture of its mum's and its dad's, but also a set of unique mutations. People, on average, carry about 60 of these unique mutations. Birds probably have fewer than this, as the avian genome is smaller than the mammalian one. Most of these new or *de novo* mutations have no obvious effect on the animal's appearance or its bodily functions – they are considered to be neutral mutations. Once in a while, though, something will occur that produces a distinct new trait, which has a definite impact on that individual's survivability. The effect may be positive or negative, and the same trait could be positive in one habitat, neutral in another, and negative in another. The most famous example of this concerns the peppered moth, a big British moth species which is normally light mottled grey for camouflage against lichen-clad tree trunks. Occasionally, dark-coloured (melanistic) forms appear through spontaneous mutation. Historically, they wouldn't survive long because they had no camouflage. However, when industrial pollution blackened tree trunks and walls in urban areas, darker mutants had the camouflage and the 'normals' did not, so in these areas the melanistic gene spread through the population and dark moths soon outnumbered pale ones.

So, genetic variability produces a mixed bag of individual animals within a population – some better adapted to survive and procreate than others. Natural selection decides which animals get to breed successfully, and the result is a gradual change in the general genetic make-up of the population, with animals becoming better adapted to survive in their environment (in other words, evolution). If environments don't change much, animals often don't change much either. If environments change too quickly, in ways that make it difficult for the animals to survive as they are, adaptation through these evolutionary mechanisms often can't keep up, and populations will shrink and perhaps die out. You can't rapidly grow better wings or develop the ability to produce more babies just because you 'need' to – new traits arrive primarily from genetic mutations, so through luck rather than design. The vast majority of living things that are affected by human-caused environmental change aren't adapting to the new way of things quickly enough (if at all) and that's why they are dying out.

Inheritable variation in behaviour can be genetic, but it doesn't have to be. Animals that are a) pretty intelligent and b) social can also pass on useful learned behaviours through 'cultural transmission' – in other words, they can learn from observing each other, and in the end the whole population has learned the new behaviour. One of the most famous examples of this concerns the Japanese macaques that live on Kōjima island. Researchers have been studying these wild monkeys for years, and have noted many examples of innovative 'better' behaviour spreading through the population, such as keeping warm in winter by lounging around in hot springs, and learning to wash sand off their food in the sea rather than brushing it off with their hands. Direct research of such cultural transmission in gulls seems a bit thin on the ground, but it's self-evident that something like the habit of tearing into tied-up plastic rubbish bags is something that one gull could learn from watching another, given their habit of quickly converging on every feeding opportunity.

There is another factor that affects how animals adapt to their environment. Time to dip a tentative toe into the new, exciting and confusing world of epigenetics. This, says the dictionary, is 'the study of changes in organisms caused by modification of gene expression, rather than alteration of the genetic code itself'. In other (and hopefully less scary) words, epigenetics deals with changes in the ways genes work, rather than structural changes to the actual DNA that makes up the genes. As a single fertilised egg divides into two cells, then four, then to a ball of cells, and eventually to a fully formed embryo with all its different parts, not every gene in every cell is 'switched on' – some are inactive. However, various processes can turn a gene from off to on, and cause the cell to make some proteins and not others. This is how all those first few undifferentiated cells are modified into different types as they divide and multiply – becoming skin cells, muscle cells, nerve cells, and so on. Epigenetics is also what causes effects like muscle cells becoming more efficient over time when we increase our exercise rate and intensity. It was always thought that epigenetic changes could not be inherited – that gene expression was 'reset to the default' in egg and sperm cells. But there is now evidence, from studies on various lab animals, that some epigenetic changes can be carried in egg and sperm cells, and so can be passed on to the next generation.

So we definitely have two pathways through which animal populations adapt – evolution by natural selection, and cultural transmission – and there may be a third too – epigenetics. What all this means is that animals adapt – a lot. And what it means for the shared future of gulls and humans is that gulls will keep finding and sharing new ways to exploit us, and if those things annoy us then we will keep finding new ways to thwart them. Down in Hastings, some of the households have been supplied with reusable, peck-and-tear-resistant 'seagull-proof' sacks for their rubbish bags. Deterrents like bird spikes and netting are widely used to discourage gulls from nesting on rooftops. And on seaside town beaches it's now very common to see signs everywhere urging people to not feed the gulls.

I was in Weymouth, Dorset, not long ago, on a sunny Sunday afternoon. I'm fond of this town, not least because it has two RSPB reserves in it – Radipole Lake and Lodmoor, both of which are really good for seeing gulls, among other things. But on this occasion we just explored the beach and harbour, and while there were plenty of gulls in the harbour, the pretty, sandy beach was almost devoid of larid life. What it had instead was a lot of strongly worded signage urging people not to feed the gulls, even accidentally. Beneath an image of a gull with a 'no entry' symbol superimposed on it was a bullet-point list of the problems caused by gull-feeding – aggression, destruction, disturbance, disease. 'Please help protect your community's health', the signs concluded, and the absolute dearth of gulls on the beach was testament to their effectiveness. Just a few carrion crows picked at the strandline in between the happy, un-gull-molested human families.

These signs and posters in Weymouth were part of a campaign first launched in May 2017, which was backed by many food-selling businesses in the town. Other anti-gull signs I've seen around include a bright red one proclaiming, 'Ice cream and chip thieves operate in this area – shield your food' with an impressionistic picture of a very gangster-like gull. Another, taking a different tack, had a photo of a quite cute-looking gull with the message 'Please don't feed me', and argued that a chip and ice cream diet can affect the birds' health. A magnificently dramatic one shows a full-frontal, gaping-mouthed gull looking horrifying, surmounting the scientifically dubious message 'Please don't feed the gulls – they are not hungry, they are just greedy'.

There's plenty of evidence that campaigns like this do work. Not on everyone of course – lots of people will still rebelliously chuck their unwanted scraps at waiting gulls, but many of us have a strong obedient streak and wouldn't dream of contravening a strongly worded sign. But food offered on purpose isn't the whole story. Even the obedient types don't tend to see accidentally dropped food as litter and will leave it where it falls, and overflowing street bins are easily accessed by gulls. Challenging that sort of thing is much harder, especially because there aren't immediate negative consequences from it. Much of the (literal) street food that sustains our seaside town gulls is provided by out-of-towners on short trips, who don't experience the mixed blessings of living among gulls all day, all year.

In my experience, people who grow up inland and settle at the seaside later in life can be less tolerant of the ways of gulls than those who've lived with them for a lifetime. This is anecdotal but I think it makes sense – you tend to retain an affection for the things of your childhood (unless they are utterly horrible, and gulls usually aren't that bad). But someone who chooses to move to the seaside for the first time in midlife may well find the gulls a lot less charming as everyday neighbours than as part of the holiday scenery. There are exceptions, of course – plenty of people moving down to the coast for the first time have told me that they love the gulls every bit as much as they love the sea air.

The most anti-gull folks of all (again, in my experience) are those who don't live in seaside towns but nevertheless have found themselves living alongside gulls. That's understandable to me. Take Liskeard, for example – the town where a pet tortoise was killed by gulls in 2015 (as I recount in chapter 5). This Cornish town is 10km (as the gull flies) from the nearest bit of coast, and 12km if you drive. It's not somewhere you'd choose to live if your priority was being right by the seaside, so I can well imagine that the tortoise owners are far from the only Liskeard residents who aren't best pleased that it is now home to more than 150 pairs of rooftop-nesting gulls, after the first pair nested there in 1994. The extremely posh city of Bath, in Somerset, is more than 30km inland and is home to a large, growing and increasingly problematic rooftop gull population, as are the nearby towns of Wells and Yeovil.

Over here in Sevenoaks, west Kent, no one really seems to have noticed (yet) that there's the beginnings of an insidious herring gull takeover underway. But I predict it won't be long before the most tempting rooftops are festooned with 'seagull spikes' and/or netting, nor before the first active (and therefore legally protected) nest mysteriously disappears.

Local councils have tried, and are trying, various other ways to discourage and scare away gulls, and 'pest control' companies offer other services in addition to spikes and netting. One of the more popular ones is the use of falconry birds to scare off the gulls. *The Evening Standard* reported on one such bird in 2018 – a harris hawk called Stella, who is flown around various sites in central London by her handler, a falconer working for Rentokil. The Harris hawk is a large bird of prey native to the Americas, and is very popular among both professional and amateur falconers because of its intelligence and easy-going nature. It is a cooperative hunter naturally in the wild, living in groups that work together to drive prey into the open. This side of its character makes it easy to handle, and it's also big and agile enough to intimidate even tough birds like gulls.

Other methods that are used to scare off or discourage gulls include devices that make loud bangs at random intervals or play gull alarm calls; aerial scarers like shiny kites, balloons and spinners; using dogs to patrol beaches where gulls congregate; electrified gadgets to deliver mild shocks to birds that go near them, and fixing lengths of taut wires along rooftop edges to prevent them from landing.

All these 'solutions' are touted as 'harmless', because they don't involve actually killing the gulls. However, they do involve harassing and stressing them to the point where they may struggle to have enough time to feed or rest, and by reducing the numbers of safe places for them to live (and we know that wilder places are not necessarily safe at all – see chapter 4). The methods that involve disrupting breeding activity are potentially more damaging, especially if they are used close to the start of nesting time. Gulls that have invested some time and effort into finding and defending a territory will not readily abandon the site, even if they have not yet begun to build their nest. Displaced pairs may not have the time or energy to make a second attempt elsewhere – and if they do not then that's the whole breeding season wasted for them. It is easy to assume that birds, when scared away from location A, will simply go to location B and carry on exactly as before, but as productive habitats shrink away, it's becoming more and more difficult for them to find new territory and new places to forage. Some of the gulls displaced from their nest sites won't be able to breed again – perhaps not ever – and some of those moved on from their feeding grounds will starve.

The predictable (and, many would say, sane) response to that is, 'Who cares, they're only gulls'. Well, this is not an unreasonable or surprising attitude. We're animals after all, and like any animal we prioritise the interests of our own species above those of all others. But we are also, now, incredibly overpowered and world-changing animals. This very natural attitude has now cost us 60% of wildlife populations around the world since 1970, and it threatens our own survival too, so we need to do something about that attitude. We definitely can't afford to keep wildlife out of 'our' spaces – our farms, our plantations, our reservoirs, our industrial areas and every other bit of the world that we use in some way. I think it is probably reasonable to keep our indoor spaces more or less wildlife-free, but as soon as you step into the open air, then as far as I'm concerned that space – wherever it is – should belong as much to wildlife as to people, as far as possible. This is what we need to do and how we need to think if we're going to do anything about the man-made mass extinction that's happening everywhere in the world. And why not start with gulls? In fact, why not start with the herring gull, complete with all its annoying ways and habits, and see if we can live happily alongside it in our towns and cities?

I grew up in a house with a herring gull nest on the roof, and while we certainly noticed they were there, we didn't have serious problems with them. The same goes for lots of householders in seaside towns. The urban environment can support a herring gull population without the birds necessarily causing all kinds of trouble, if we are a little more tolerant and a bit more sensible in our habits. Gulls aren't going to stop shrieking nor voiding their bowels, so we have to learn to deal with a certain level of noise and mess. But we can 'train' them to leave our bins alone, and not to dive-bomb us.

The first thing to do is be careful about food. The herring gull is a total omnivore and any food left outdoors, in the open or with only a flimsy covering (like a plastic bag) will be located and eaten. Gulls, unlike most birds, have quite an acute sense of smell, and can use this to track down food (as well as to find their way on migration, as we saw in Chapter 3). So – don't drop or leave litter, don't stuff more litter into street bins that are already full to overflowing, and don't leave your bin bags out overnight but put them out in the morning on bin day, because gulls get up much earlier than the bin collectors do. Councils in areas with gulls (or foxes, or other wildlife that knows what might be in a bin bag and how to get at it) are slowly learning that people *do* still put out their bin bags the night before, because sometimes there's no choice, and that wildlife-proof containers need to be provided. These might be wheelie bins, or peck/bite-proof sacks – whatever they are, they will solve the problem, at least until gulls work out how to use power tools.

We also need to stop feeding the gulls on purpose. This is very, very tempting, because feeding gulls is fun and it makes us feel good generally to feed the birds. So I fully sympathise if you're a seaside dweller who's in the habit of tossing your spare chips to the laridine hordes. I also confess to hypocrisy, because last Tuesday I was waiting for a late train on platform three at Hastings station, and diverted myself by throwing a few cheesy Wotsits to a beautiful herring gull who'd come to sit beside me (she didn't drop a single one). So I have to include myself in that 'we', and try to resist the next hungry-looking larid that comes along to scrounge my crisps. Charming though gulls can be when they throng around, and impressive though they are in their agility, feeding them like this will whittle away their fear of humans. The end result is what we see in some seaside towns, such as St Ives in Cornwall – a population of gulls that think nothing of grabbing food directly from your hand, or mouth, or jumping up and down on your head to try to persuade you to hand them a snack. Habituating all but the tiniest wild animals too much can lead to scary and sometimes painful encounters, and that naturally leads to animosity. Ultimately it's the wild animals themselves that suffer when humans decide to hit back.

Feeding birds in the town park is another matter. Even staunch gull-dislikers don't tend to mind seeing a few gulls bumping shoulders with the ducks and moorhens on the park pond, and there's pretty much nothing you can offer the ducks that the gulls won't also eat. It will, however, benefit all the parkland birds if people refrain from chucking armfuls of bread into the water as the birds won't manage to eat it all, and it's not terribly good for them anyway. Uneaten bread encourages algal blooms, which is bad for all the pond wildlife, so it's best to feed the park birds in moderation. Pellets of floating duck food are easier than bread for the ducks to eat, being suited to their dabbling/nibbling bill action, and are less likely to encourage the gulls, which would rather grab and wolf down an entire crust of sliced white.

Reducing the food supply for urban gulls in these ways will reduce the gull population to what can be sustained by natural food, some of which we actively want the gulls to be eating. They are, after all, scavengers in the main – nature's clean-up crew. They will tidy up washed-up dead things on the strandline and consume roadkill, as well as catching their own prey.

What about gulls nesting on rooftops? We've seen in Chapter 5 that they can (occasionally) be seriously problematic once their chicks hatch, becoming over-protective. If you live in a bungalow or other low-profile house, it's probably wise to put up spikes or otherwise modify your rooftop to make it gull-unfriendly (outside the breeding season, so you don't disrupt a nesting attempt and break the law into the bargain), but gulls nesting on taller houses are very unlikely to harass you when you go outside.

That leaves you with the issues of noise and mess, and talking about this reminds me of something I read on an internet forum a few years ago. A woman on this forum started a thread asking for advice about birds in her garden. She didn't like them. Note that this wasn't gulls, nor crows or vultures, nor any other generally unpopular collection of feathered menaces, but the usual little songbirds that frequent your average British back yard – sparrows, finches, tits, robins. They spent a lot of time in a particular tree in her garden and their twitterings and pooings were driving her mad – was there any way she could prevent them from perching in this tree? I replied to the thread, suggesting she uproot the tree and move it into her house. It was my way of saying what I already said in this chapter – the outdoors should have wildlife in it, lots of wildlife. And we should be willing to put up with the small inconveniences that this can bring, for the privilege of sharing this space and this world with such a wonderful array of other living things. Charity begins at home and so does conservation. So if your local gulls are pooing on your car, park elsewhere or drape something over it while it's parked. If their yowks and yelps are waking you up, wear earplugs.

It turns out that it's not that easy. People carry on disliking gulls and the things that gulls do, and many people feel that they have – or should have – the right to live in gull-less towns. Even today, when none of us can really have escaped the knowledge that our wildlife as a whole is in absolute crisis, we still have selective blindness when it comes to certain categories of wildlife, especially in urban areas. That means it's time to unveil the super-weapon in this battle. Grab your binoculars and go down to the coast, and look at some gulls. If you don't have any binoculars, it probably doesn't matter. Seaside gulls are usually pretty approachable. Look at them, watch them, notice what they do, their interactions, how they sound, the way they fly, how their feathers fall into place when they settle. Check out how many individual differences they show, even within the same species. Notice how bonded pairs demonstrate their affection for one another. Consider that some of those gulls could be in their twenties or thirties, that some of the pairs you see could have been coupled up for a decade or more.

Familiarity, when it's unforced like this, doesn't breed contempt but admiration (that's my own experience and my hope for others, anyway). Watching kittiwakes skimming a stormy sea, their translucent flight feathers flaring silver-gold when the sun shines through them, is one of those throat-tightening, uplifting moments, but looking into the clear cool eye of a herring gull at point-blank range could be as well. It's really difficult not to be overawed by the brute power of a great black-back, nor charmed by how tenderly a parent black-headed gull serves up a semi-digested meal for its adorable fluffy nestlings. Think of all these birds, leading their lives for so many millennia, and now fighting to survive in a world that's changing at breakneck speed. Surely we can cut them some slack, and concede them some space?

When I was a child I used to hate the way that every natural history TV programme would, in its last ten minutes, go into 'guilt-trip mode'. The worst of it was that sense of helplessness. How can I, personally, stop people in India killing tigers for their body parts, or end the war that threatened the last vestiges of mountain gorilla habitat in east Africa? It's not that different today, except that the dire warnings are much more wide-reaching. No longer one species and one issue at a time – everything natural is going to ruin, wholesale. Our seas are awash with plastic waste, our ice caps are melting, all of our wild spaces are being razed and built upon. The things each of us can do as individuals to reduce our damaging impact are drops in the ocean, grains of sand in the desert. And some of the solutions that we're considering now are for our benefit alone – they will only make things worse for wildlife. Crop production threatened by the loss of pollinating insects? Don't save the bees, build robot ones instead! Then we can keep using pesticides to kill off all the non-helpful insects! We might *have* to do things like this someday, but even robot bees won't be enough to keep us alive on a dying world in the end.

Science fiction scenarios would eventually see us (or at least the richest 0.0001% of us) eventually packing up and leaving Earth behind, to start again on a new world and hopefully do better this time. But that's much less realistic than any hopes of slowing and reversing the damage we're doing before it makes our world uninhabitable. We don't need escapism, we need optimism, and determination, and compassion. This is where gulls matter, along with all the other organisms that are (so far) making a go of survival in the most non-natural places on this drastically human-altered planet, alongside the rats and cockroaches, the crows, foxes, silverfish, pigeons and peregrines, mice and houseflies. Their presence helps support the presence of other, less obvious strands in the ecological web that we all rely upon. And they keep us in touch with wildness. They remind us that where there's life, there's hope.

CHAPTER 8

2019 – A Gull Postscript

Black-headed Gull

YESTERDAY – 2 DECEMBER 2019 – I was having my usual morning rummage through social media, and happened upon a set of photos of a Ross's gull. The pics were taken in the USA, by the edge of Lake Washington. That's a long way out of range for this Arctic gem – in fact this was only the third state record. I would have been riveted by the photos anyway. Such an exquisite bird – pint-sized, wide-eyed, tiny-billed, pink-flushed and silver with a pair of crazily long, blade-sharp wings that drooped at its sides, as if tired out from flying far too long. I read the story of how the rarity was discovered, and was watched by a fast-growing crowd of delighted birders for a couple of hours. Then it flew out to the lake, where it was promptly caught and carried away by a passing bald eagle. I hoped that there'd be some happier gull stories to read before the year was out.

This has been a year of almost paralysing uncertainty and frustration for those of us living in the UK. Yet the newspapers still had their usual 'silly season' in summer, where a high proportion of news stories were what you might call 'local interest' rather than 'national misery'. The silly season of 2019 featured a very hefty dose of gull-based headlines, and in some quarters anti-gull sentiment was ramped up to unprecedented levels. More about that later. For me, though, it was momentous in another way, as I finally returned to the seaside.

In July, I moved to Hythe, a small and mild-mannered seaside town close to Folkestone and just a short gull's flight along the coast from Hastings, where I grew up. So for me, it's been like coming home, in lots of ways. If I should feel the need, I can walk to the seafront, sit on a bank of shingle, watch the waves wash in and out with their escort of gulls, and feel my mind settle into quiet and calm. When I leave the house, if I turn around when I'm a few paces away I'll probably see 'my' herring gull perched proudly on the chimney, casting her cool gaze across the street and the town beyond. On the hottest days in summer, when the flying ants climbed up our walls and launched themselves into the air for a frenzy of lovemaking, the blue skies became more gull-filled than ever. I took my camera out to photograph them, because most of those gorging themselves on the ant bounty were Mediterranean gulls. Ten minutes of shutter clicking and I had my lifetime-best photos of Meds in first-summer, second-summer and adult summer plumage, and best of all, a couple of stunning juveniles – fledglings of the year, perfect to the last brand-new feather. Not even a gull with the world's largest stolen chip could have been as happy as I was that day.

But what of the PR disaster that befell gull-kind in 2019? The most notable of the gull-themed news stories was one to horrify any (small) pet owner. On July 22, 2019, the BBC informed us that 'a seagull has seized and flown off with a family's pet Chihuahua, according to its owners.' The dog, a four-year-old, 2kg male called Gizmo, was abducted from a garden in Paignton, Devon. The bird seized its victim by the neck, and the owner's partner tried to grab the dog's dangling legs but the gull had already risen too high. The dog was never seen again – which led to further speculation that the gull must have swallowed the dog whole. A leg bone was then found on a rooftop and analysed, but was found to have belonged to a rabbit.

Various experts, asked for their views, said that such an occurrence, though surely very rare, was not outwith the realms of possibility and that owners of small pets should be careful about letting them roam around outside. Various other experts, though, said nope, this almost certainly could not have happened – for reasons I'm going into now.

I've already presented a few cases of gull-on-pet violence, but this is different. A gull going for a mammal on the ground is one thing. A biggish adult rabbit weighs about 2kg – the same as the chihuahua in the story. If a large gull (say, a male herring gull, weighing in at about 1.1kg) found a 2kg rabbit and for some reason the rabbit could not flee, I would absolutely expect the gull to laboriously kill and dismember it with repeated bill strikes and eat it. Gulls aren't raptors – they don't have the physical equipment to easily deal with prey twice their own weight, but what they lack in efficiency they make up for in enthusiasm. But could that gull lift that live rabbit bodily into the air, even as it struggled as best it could, and fly away with it?

Owls and raptors can carry heavy weights but they do it with their feet – something a web-footed gull could never do. Their sharp curved talons sink in and the weight is borne by the full muscular power of their whole bodies. Could a neck take that same strain, while still being mobile enough to allow the quick head movements that are sometimes necessary in flight? And how much strain are we talking, anyway? It's widely known that even the mightiest hawk or owl cannot carry off something heavier than itself – it can *kill* said prey but can't take it away. This is one of the reasons why these birds can tear into a kill so quickly and eat lots of meat in a short space of time – if they've killed something that's too heavy to carry to a safer place, they have to eat it on the ground and inevitably there will soon be scavengers around to try to steal the kill.

What if it wasn't a herring gull but a great black-back? These gulls do spend time around towns and might come down into gardens, though much less often than herring gulls. And anyone who's been subjected to the steely gaze of the king of gulls at close range would testify to its size, and the size of its bill, but even the biggest examples of this gull don't hit 2kg – the heaviest one recorded in Britain was

1.92kg. Even the most mighty-looking of flying birds are lightweights compared to similar-sized mammals – their bulk is, literally, feather weight. I did find a single photo online of a great black-back in flight carrying a rabbit. It was a baby rabbit – by its size relative to the bird, I would guesstimate its weight at about 0.5kg. It was being carried downwards, not up (from a cliff-top down to the sea) and even so it looked hard work for the gull.

Contemplating the maths, I'm reminded here of the bit in the film *Monty Python and the Holy Grail*, where King Arthur and a castle guard have a heated debate about whether a swallow, or two swallows, could fly while carrying a coconut. I'm also reminded of the number of times I've been asked whether dorsal guiding feathers are a real part of avian anatomy (they're not). The sceptical guard conclusively won that argument, and the same physics says that the gull–chihuahua story could not have happened, at least not in the way that it was told. But physics can surprise us, and so can gulls – so the truth has to remain a mystery. What did definitely happen is that a much-loved pet was lost, and a lot of people became very angry and very paranoid about gulls – though, of course, a lot of people already felt that way.

So we have learned that gulls may be even worse than we thought. We already knew they were pet murderers but perhaps now they are also pet abductors. And as news stories about wildlife extinctions grow ever more world-encompassing and depressing, our personal tolerance for the actual wildlife that surrounds us seems to grow smaller and meaner by the year, and our actions to try to suppress it (for all sorts of short-sighted and/or spurious reasons) grow more and more bizarre and wrong-headed.

In early spring 2019, many people were baffled to see their local hedges and trees covered in mesh netting, placed there by local council workers to discourage wild birds from nesting. Why would anyone do this? It remains illegal to destroy or interfere with any active bird's nest in this country, so developers who want to chop back hedges or cut down trees during the breeding season are putting up the nets to get around this law – no chance of contravening it if there are no nests to begin with.

Along similar lines, when migratory sand martins returned to their seaside nesting colony at Bacton in Norfolk, they found their burrows blocked off with mesh too, the council having decided that the nest-holes in the sandstone cliffs might increase the risk of coastal erosion. Maybe the council expected the birds to fly off immediately and try to find another site to breed, but that's not the nature of sand martins. For days, the distressed birds flew around and tried to get through the mesh, until finally furious protests from wildlife lovers convinced the council to relent and let the birds access their nests – thankfully just in time to allow a normal breeding season to take place.

Gulls have, of course, also been targeted by some, frankly, bonkers initiatives. Another 2019 summer story from the BBC – at the popular swimming beach in Worthing, just along the coast from Brighton, locals observed falconers flying hawks over the sea in a bid to drive away gulls. Southern Water, the company who arranged it, said that this 'vital work' was to prevent the gulls from polluting the seawater by defaecating into it. I confess that I feel quite inadequate as a writer when I try to articulate my feelings about this one. I suppose the hawks must have been trained not to make matters worse by pooing into the sea themselves. Perhaps Southern Water (who, incidentally, were fined £126 million in 2019 for allowing a large amount of untreated waste water to enter the environment from their sewage plants) will also be introducing some similarly well-trained sharks into the bathing waters to discourage the many fish which might also pollute the seawater with their waste.

Here are a couple of other 2019 news headlines that jumped out at me. 'England could have its first seagull cull in 40 years as city considers application to Natural England' (*The Telegraph*), and 'Devon's seagull cull: The day GUNSHOTS sounded across the bay' (www.devonlive.com).

Just to clarify, these two are not related. The city that was considering applying to Natural England for a licence to shoot some of their breeding lesser black-backed gulls is Worcester – in the West Midlands. The city council's environment committee decided not to apply for the licence after being advised that they stood no chance of success, but they are implementing other, less illegal measures. These include using drones to locate active nests on rooftops, and then climbing up to the nest, removing the gulls' eggs and replacing them with dummy versions. I don't know how much longer a gull pair would tend a nest of dummy eggs after their 'due date' for hatching had passed, but gulls only make one breeding attempt per year, so this tactic could wipe out the best part of a year's whole cohort of youngsters.

Although the Devon Live headline comes from 2019, it actually concerns local people's recollections of a day in 1975, when an organised gull cull took place in Torquay. Gunshots did indeed sound across the bay, there was a tremendous outcry from the locals, and just six years later the Wildlife and Countryside Act put a stop to such events – for now, at least. Every year, there are loud clamours for gull culls from all quarters, including opinion pieces on Devon Live's website. The website helped to fan the flames a little more with another 2019 story, startlingly headlined 'Seagulls drunk on beer and ants are falling over and vomiting: the cocktail is reported to be making them more anti-social'. The article speculated that gulls were getting drunk by draining half-full plastic beer cups left on the streets by revellers, or possibly eating some kind of brewing by-product, of unknown provenance. They were then making everything much worse by consuming vast amounts of flying ants, and this triggered wild hallucinations because ants are full of formic acid.

This story definitely warranted a bit of investigation. I have photographic evidence of gulls drinking from abandoned beer glasses – it's not an unusual sight on Hastings seafront early on a Sunday morning. And if some kind of tasty, alcohol-laced brewing by-product was left lying about I have no doubt that gulls would eat it. I have not personally observed any gull drunkenness – but it's well established that alcohol has a similar affect on birds as it does on people, and if gulls were getting hold of it, by whatever means, then they would be getting drunk. The article quotes several RSPCA staff, one of whom describes, rather vividly, a call-out to deal with a drunk gull that had fallen off a roof, and had then proceeded to douse its rescuer with strongly beer-scented vomit.

What about the hallucinations? Well, as I mentioned earlier, gulls do indeed love scoffing flying ants. But could eating too many flying ants really send them off on some kind of hallucinatory adventure? I found out that formic acid, though it can make you go blind, isn't a hallucinogen. But then I stumbled upon an account of how certain indigenous people in California induce a hallucinatory state by munching handfuls of red harvester ants, wrapped up in eagle feathers. This 1996 account went on to explain that, while no hallucinogenic substance has yet been isolated from the body of any insect, the ants' venom did contain certain chemical compounds which were 'pharmacologically interesting'.

Perhaps seeing gulls at the mercy of the demon drink and other narcotics will help us to empathise with them more, rather than demonise them. And maybe there is hope that we can find new and non-harmful ways to deter them when they do become a nuisance. According to one of my favourite 2019 BBC stories, 'staring at seagulls helps protect food, say scientists'. The research, carried out by scientists at the University of Essex, entailed recording how long it took gulls to approach dropped chips when a person nearby was staring at them, and when the same person (at the same distance) was ignoring them. On average, the gulls took 21 seconds longer to grab the chip when being scrutinised. Many local seaside-town newspapers picked up the story, and *The Argus* in Brighton even recreated the experiment and filmed it. The journalist placed a chip on the pavement near some interested-looking gulls and then went into a low squat with either his face or his bum hovering over the chip. The results make convincing (and amusing) viewing. It is well known that a direct human stare disconcerts many animals. They know what eyes are and they know what it means to be looked at. Our eyes face forward, a trait shared with many predators, and we stare when we want to assess distance – in other words, we probably look (to the gull) like we're about to pounce.

In 2019, the much-revered ornithological journal *British Birds* also took a look at urban gulls, with a rather more considered eye than the tabloids. Sarah Trotter's article 'The regulation of urban gulls in the UK: a study of control measures' looked at the various ways that local authorities (LAs) had dealt with (or tried to deal with) their perceived gull problems between 2010 and 2016. The methods examined fell into two broad categories – either they targeted the gulls themselves, or they took the form of advice or directions given to people in relation to gulls. The former category includes installing bird-proofing devices (used by 19 LAs), direct removal of nests or eggs under the General Licence (16 LAs) and use of falconry birds to scare gulls away from their nesting sites (eight LAs). In the latter category, 41 LAs produced posters and leaflets advising the public not to feed gulls, and 12 issued Community Protection Notices to individuals whose bird-feeding habits (though they were not necessarily feeding gulls) had been reported by other local people as having a detrimental effect on their quality of life. In total, the 74 LAs that responded to the author's query described using nine different anti-gull measures, and seven different human-targeted methods. The LAs spent as much as £87,000 a year on this work, but many felt that they were yet to solve the problem. Sarah Trotter's article ends with an appeal to adopt a different way of thinking about gulls – approaching them purely and simply as 'urban pests' offers no space for the chance of peaceful coexistence. Yet peaceful coexistence with wildlife in general is ever more urgently needed.

The year 2019 has seen the rise of Extinction Rebellion, a protest group that has formed in response to ever more unignorable evidence that we are driving our world's wonderful ecosystems to absolute destruction. Civil disobedience does help get things done – would women have the right to vote without it? But here in the UK, in the December 2019 general election, women and men have just voted back into power the party with the least environmentally friendly manifesto of all, according to an assessment by Friends of the Earth. Out of a possible 45 policy points, Labour scored 33, the Green Party 31, the Liberal Democrats 30, and the Conservative Party just 5.5. Even when we have the chance to do it, many of us are not yet backing action on climate change and other environmental issues, and it's difficult to imagine what more has to happen before we will.

Despite all of those news stories, I am as guilty as the next UK-dwelling person of not *feeling* it all as strongly as I should – it still doesn't quite seem completely real, or immediate. But there's something about living by the sea that does makes it all feel a little more of both of those things. A few weeks ago, the weather was stormy and we in east Kent were told – by social media, news headlines and even an actual phone call to our house – that a big and dangerous tidal surge was on the way the following morning. The Met Office warning map showed a long stretch of coast between Folkestone and Dungeness as under threat – the self-same bit of coast that we are likely to lose to the sea over the next fifty years if things don't change.

The day dawned wild and I was apprehensive. We had errands to run down the road in Folkestone. On the way back along the seaside as the tide rose to its peak, we watched several hardy souls completing the Folkestone half marathon, ducking the gouts of sea spray that were drenching the promenade. Later, when it became apparent that we weren't actually going to be washed away (this time), we headed west for a walk along the shingle at Dungeness. The waves towered, roiled and smashed their way in. Great black-backed gulls tipped and skimmed over the white horses close in to shore, while further out a couple of young arctic skuas – dark, dashing and exuberantly energetic – chased after the Sandwich terns. On the beach we found a plastic jar with its lid covered in big, glossy-shelled goose barnacles – fascinating stowaway crustaceans that voyage the oceans when they attach themselves to flotsam rather than rock. The jar didn't have any markings that would have made it easy to tell where it came from, but often goose barnacles wash ashore attached to objects that come from far-flung shores. One cluster, found on the Isles of Scilly, was attached to part of a spacecraft that had originally been launched from Florida. The sea shrinks our world; highlights its frailty.

Sometimes it's depressingly easy to fill a bag with all kinds of discarded or washed-up plastic debris when you make a beach visit. Sometimes, it's not, but that's largely down to the weekend efforts of dedicated crews of volunteer litter pickers. And, if it weren't for man-made sea defences, there wouldn't be a beach here at all. Here in Hythe we have a couple of mighty breakwaters, made from huge boulders. They provide foraging grounds for turnstones and purple sandpipers in winter, and shelter for swimmers in summer, but even they aren't enough to keep this beach safe from the action of longshore drift. The sea is constantly taking away the shingle from here and depositing it further east, so every now and again, trucks carry tonnes of shingle back to Hythe. It's a rather blunt-instrument approach to coastal defence, and one that may need a drastic rethink in the not-too-distant future. But we humans have invested too much in our coastal towns to allow the sea to reshape the coastline in its natural way, so we will be fighting this war for a long time to come, one way or another. And we'll be reminded ever more pressingly of the power that we struggle against, as climate change does its work and the sea level steadily rises.

It is now time for me to write a postscript to this postscript, because it's March 2020 and planet Earth has become a rather different place for a large swathe of the almost 8 billion humans that live here. We're not talking about summer holidays, Brexit or even climate change very much any more. Instead we have a new lexicon. Pandemic. Social distancing. Lock-down. Self-isolation.

At the very end of 2019, a new and deadly strain of virus was identified, in patients in Wuhan, China. This virus, causing a disease known as Covid-19, is a coronavirus (which means it is sphere-shaped with an all-over 'crown' of spiky glycoproteins), and it may have infected its first victims via wild bats sold at 'wet markets' in parts of China. However it arrived in the human population, it has spread at an extremely alarming pace, thanks to its highly infectious nature (and long incubation period, meaning that many who spread it have no idea they are already infected), and the prevalence of global travel.

The first confirmed cases in the UK were diagnosed on 31 January 2020. Today, just over 17,000 cases have been recognised and just over 1,000 people have died, the first of them on 5 March. Obtaining an accurate picture of the death rate is close to impossible as most people with symptoms are not tested, so the number of reported cases is made up primarily of those whose symptoms have been severe enough to require hospitalisation. It is clear that, in younger and healthy patients, Covid-19 is often not a serious illness and no hospital treatment is needed. Nevertheless, with a population of almost 68 million, and more than 15% of them over 60 years old (and plenty more suffering from health conditions that make them more vulnerable), the total number of extra deaths from uncontrolled Covid-19 infection is potentially very high.

As I write these words, on 29 March 2020, 199 countries around the world have recorded cases of Covid-19. In some countries, particularly in Europe, the spread has been so rapid that radical 'lock-down' measures have been implemented, to try to control the virus's spread. We will have no vaccine for many months, but the hope is that the virus's rampant march through our lands can be held back in the meantime, at least to the point that national health services are not overwhelmed by the scale of the emergency. Here in the UK, we have been locked down for six days and will remain so for at least two more weeks (and very probably much longer).

The strictness of the lock-down varies between countries. Here, we are strongly encouraged to stay indoors as much as possible, and can be fined if we are out without good reason. The reasons include shopping for essentials, attending medical appointments, working (only if your work is vital and working from home is impossible), exercising ourselves or our dogs, and helping a person in need. In effect, no-one is really going anywhere any more. Restaurants, pubs, theatres and non-essential shops are closed, as are schools, and numerous offices and other workplaces. Outdoor gatherings are banned. When out on our essential errands, we must stay two metres away from any other humans we meet, at all times. Driving is not encouraged, because each car journey carries the risk of an accident that would further stretch the health services, and each human hand on a petrol pump could spread the coronavirus.

This strange time brings an array of challenges for human beings. We have to stay in, and it's hard to ignore the news stories and the death-rate graphs that show a curve steeper than the north face of Everest. A great many people have lost their jobs. For a few of us (so far), there have been untimely bereavements. For many of us, lock-down means extreme isolation, which severely tests our mental health, and some are locked into dangerous situations. For nearly all of us, even if all is generally well in our lives the whole experience is a deeply unsettling combination of boring and terrifying. But there are good feelings to be found as well in the midst of the panic, sorrow and worry. It's a time to appreciate, beyond measure, the workers who are still going out into this suddenly dangerous world and risking their necks to provide health care, food, and vital services. It has also been turning into a time to appreciate nature a little more than before.

Outside, the wind's whipping up the trees and stroboscopic sunshine flicks on and off between the dashing banks of cloud. The herring gulls, comfy with this degree of aerial turbulence, just hang there on still wings and let themselves be lifted and drifted at random. There seem to be as many about as ever. It is time for pairs to reaffirm their bonds and renovate their nest-sites. No lock-down for them, but what effect will a sudden absence of humans in the outdoor world have upon them?

Air pollution in many parts of the world affected by lock-downs has shown a dramatic drop, particularly in cities. Nitrogen dioxide levels in particular have fallen greatly – by more than 50% in just a week in Madrid, for example. In the UK, comparing air quality between 1 January–10 February against 15 February–25 March shows that nitrogen dioxide and 'tiny particle pollution' (which includes metal dust from vehicle brake pads as well as lots of other stuff you wouldn't want to inhale) has gone down by between a quarter and a half in big cities such as London and Cardiff. This is, of course, great for wildlife as well as (ultimately) for humans, especially urban wildlife like rooftop herring gull colonies, though whether we will sustain any of our less polluting behaviours after this crisis has passed is by no means certain.

Fewer humans out and about, on foot and in their cars, is also good for wildlife. Birds that nest on the ground are less likely to lose their broods to marauding dogs or careless walkers. Wildlife roadkill will fall. There will be less potentially harmful litter in the sea, on the beaches, and everywhere else. And many stuck-at-home people, especially families with younger children, may turn their gardens into wildlife havens to give themselves something to do and to watch, and become more invested in the long-term well-being of nature. A shortage of food in the supermarkets has also encouraged a spate of enthusiasm for grow-your-own. Our gardens will become miniature, personal allotments – the lifeless concrete patio will fall from favour and the vegetable patch will hold sway.

Gulls, though, rely ever more on what humans leave behind. So they, along with the other urban commensals, might suffer more than they benefit. Already, some of the residents in my seaside town are reporting that they're seeing far fewer gulls around on the now-quiet beaches. Might a spring and summer lock-down actually harm their populations or at least threaten their breeding success? It seems very possible to me, though let's not forget that the overwhelming majority of all our breeding gulls, even herring gulls, still do their breeding well away from human settlements. These rural gull populations could benefit from the changes that are happening in this very strange year.

The news tells us that life, here in the UK, will be disrupted for at least six months. It seems an eternity but really it's a blink of an eye. The problem of climate change and everything that goes with it will still be with us when this current crisis has receded. But perhaps the advent of Covid-19 will change the way we look at the longer-term problems, too.

Among the many topical memes that have emerged of late, one that particularly struck me was a simple block of text that said 'I feel like mother nature has sent us all to our rooms, to think about what we've done.' It is almost certain that Covid-19 came to us via wild mammals – some of them endangered – that are brought alive to wet markets to be butchered and sold. There are, of course, any number of other modern-day human practices that are incredibly harmful to wildlife, but this is one of the most overtly shocking. The idea that Covid-19 is nature's way of striking back is powerful, frightening and humbling.

China have now banned the sale of wild animal products at their markets – a very significant cultural shift – and there is pressure on other countries to follow suit. The big question is whether this is the start of large-scale, meaningful action to deal with the wider problems that threaten our species' long-term survival, as well as all other life on Earth. The problems remain immense but there is reason to hope that the time has come for us to learn more, and change more, and ally ourselves with nature, in genuinely committed ways.

What remains very clear is that we are going to need our wildlife, to help get us out of this mess. We need the wild and natural world to survive and thrive if we too are to remain well – physically and mentally – and we need to learn to respect wildlife much more than we have before. We're lucky that – as far as we know – bees and butterflies, tigers and turtles, and gorillas and gulls don't appear to hold grudges.

Acknowledgements

First of all, I would like to thank Julie Dando for commissioning this book. I have worked with Julie for many years in various editorial ventures, and was delighted when she and the equally talented Marc Dando founded their own publishing company. I knew from the start that their books would be very special and I am so proud to be joining the Wild Nature Press list. I'd also like to thank Robert Kirk at Princeton University Press for publishing this book over in the land of the American herring gull, and for his helpful suggestions and feedback.

I sought the advice and views of a host of birdwatching and non-birdwatching friends and acquaintances, as well as conservationists and researchers while preparing this book – thank you to everyone who answered my questions, recounted their experiences and did their best to explain why they (mostly) love gulls so much.

Thanks again to Julie and Marc for tidying up my illustrations and creating the lovely page designs. Thanks are also due to Rowena Millar for editing my text and providing numerous helpful comments, also to Judith Gibson for proofreading. And finally, thanks to the very talented Jeremy James for producing the wonderful cover artwork.

Index